Hugh T. W. Tan
Giam Xingli

Plant Magic

Auspicious and Inauspicious Plants from Around the World

Designed by Rachel Chen
Editors: Lee Mei Lin and Yaun Marn Cheong

Published by Marshall Cavendish Editions
An imprint of Marshall Cavendish International
1 New Industrial road, Singapore 536196

Other Marshall Cavendish Offices:
Marshall Cavendish Ltd. 5th Floor, 32-38 Saffron Hill, London EC1N 8FH, UK • Marshall Cavendish Corporation. 99 White Plains Road, Tarrytown NY 10591-9001, USA • Marshall Cavendish International (Thailand) Co Ltd. 253 Asoke, 12th Flr, Sukhumvit 21 Road, Klongtoey Nua, Wattana, Bangkok 10110, Thailand • Marshall Cavendish (Malaysia) Sdn Bhd, Times Subang, Lot 46, Subang Hi-Tech Industrial Park, Batu Tiga, 40000 Shah Alam, Selangor Darul Ehsan, Malaysia

National Library Board Singapore Cataloguing in Publication Data
Tan, Hugh T. W.
Plant magic : auspicious and inauspicious plants from around the world / Hugh T.W. Tan, Giam Xingli. – Singapore : Marshall Cavendish Editions, c2008.
p. cm.
Includes bibliographical references and index.
ISBN-13 : 978-981-261-427-8

1. Plants – Identification. 2. Plants – Mythology. 3. Plants – Folklore.
I. Giam, Xingli. II. Title.

QK45.2
580 – dc22 OCN237157230

Printed in Singapore by Times Graphics Pte Ltd

To our families for all their love and support

Contents

vi List of Plants

ix Preface

x Introduction
- What are plants?
- Sacred or magical plants; plants of mythology and superstition

xii How to Use this Book

1 The Plants

183 Glossary

189 References

194 List of References

204 Index

216 Acknowledgements

List of Plants

2 *Adenium obesum* • desert rose, impala lily
4 *Aegle marmelos* • bael fruit, Indian quince
6 *Alstonia scholaris* • devil tree, blackboard tree
8 *Ananas comosus* • pineapple
10 *Azadirachta indica* • neem, nimtree
12 Bambuseae • bamboo
14 *Betula* species • birch
16 *Camellia japonica* • camellia
18 *Cananga odorata* • ylang-ylang
20 *Canavalia gladiata* • sword bean
22 *Cardamine pratensis* • cuckoo flower, lady's smock
24 *Castanospermum australe* • lucky bean
26 *Ceiba pentandra* • silk cotton tree, kapok tree
28 *Cercis siliquastrum* • Judas tree, love tree
30 Christmas Tree
36 *Cibotium barometz* • Scythian lamb, woolly fern
38 x*Citrofortunella microcarpa* • limequat, calamondin
40 *Citrus medica* var. *sarcodactylis* • Buddha's hand, Buddha's hand citron
42 *Citrus reticulata* • mandarin orange, tangerine
44 *Cocos nucifera* • coconut, coco palm
46 *Conium maculatum* • poison hemlock
48 *Cordyline fruticosa* • ti plant
50 *Crassula ovata* • jade plant, friendship tree
52 *Crataegus* species • hawthorn, thorn apple
54 *Curcuma longa* • turmeric, yellow ginger
56 *Dendranthema xgrandiflora* • chrysanthemum
58 *Dianthus caryophyllus* • carnation
60 *Dracaena fragrans* • cornstalk plant
62 *Dracaena sanderana, Dracaena sanderiana* • lucky bamboo
64 *Epiphyllum oxypetalum* • queen of the night
66 *Euphorbia pulcherrima* • poinsettia, Christmas plant
68 *Ficus benjamina* • Benjamin fig, weeping laurel
70 *Ficus carica* • common fig, cultivated fig
72 *Ficus microcarpa* • laurel fig, Malayan banyan
74 *Ficus religiosa* • bodhi tree, sacred fig tree
76 *Fortunella japonica, Fortunella margarita* • golden orange, kumquat
78 Four Leaved Clover
84 *Fuschia* species and hybrids • fuschia, lady's eardrops
86 *Gladiolus* species and hybrids • gladiolus, sword lily
88 Heavily Fruiting Plants

90 *Hedera helix* • ivy, common ivy
92 *Hordeum vulgare* • barley, malting barley
94 *Hyacinthus orientalis* • hyacinth
96 *Hypericum* species • St. John's wort
98 *Ilex* species and hybrids • holly
100 *Jasminum sambac* • Arabian jasmine
102 *Juniperus* species and hybrids • juniper
104 *Lagenaria siceraria* • bottle gourd
106 *Laurus nobilis* • bay tree, bay laurel
108 *Lilium* species and hybrids • garden lily, lily
110 *Mangifera indica* • Indian mango
112 *Michelia xalba* • white champaca, white sandalwood
114 *Michelia champaca* • champaka, orange chempaka
116 *Musa* species and hybrids • banana
118 *Nelumbo nucifera* • lotus
120 *Nycanthes arbor-tristis* • night-blooming jasmine, tree of sorrow
122 *Ocimum tenuiflorum* • holy basil, sacred basil
124 *Olea europaea* • olive
126 *Oryza sativa* • rice
128 *Pachira aquatica* • money tree, knotty pachira
130 *Paeonia* cultivars • peony
132 *Phoenix dactylifera* • date palm
134 *Pinus* species • pine
136 *Platycladus orientalis* • Chinese *arbor-vitae*, oriental thuja
138 Prickly, Spiny or Thorny Plants
142 *Prunus mume*, *Prunus salicina* • plum blossom, Japanese apricot
144 *Prunus persica* cultivars • peach, peach blossom
146 *Prunus serrulata*, *Prunus xyedoensis* • cherry blossom, Tokyo cherry
150 *Punica granatum* • pomegranate
152 *Rosa* species and hybrids
156 *Rosmarinus officinalis* • rosemary, friendship bush
158 *Saccarum officinarum* • sugarcane
160 *Salix caprea* • pussy willow
162 *Salix* species and hybrids • willow, weeping willow
164 *Saraca asoca* • asoka tree
166 *Solanum mammosum* • apple of Sodom, nipple plant
168 *Thuja occidentalis* • white cedar, American *arbor-vitae*
170 *Thysanolaena latifolia* • Asian broom grass, bamboo grass
172 *Vicia faba* • broad bean, fava bean
174 *Viscum album* • common mistletoe
176 *Vitis vinifera* • grape, common grapevine
178 *Zamioculcas zamiifolia* • zz plant, money tree
180 *Zantedeschia aethiopica* • arum lily

Preface

Plants feature very heavily in our lives and we depend on them, primarily, for food and shelter. We make paper from wood and medicines from the biochemicals in plant extracts. Beyond the countless practical uses plants can be put to, their ornamental and aesthetic contributions also reveal how integral they are to our heritage and lifestyle. History, culture, religion and traditions have, over the ages, shaped the symbolism of plants, rendering some auspicious (favourable, lucky, propitious), and others inauspicious (unfavourable, unlucky, unpropitious).

This book is an attempt to catalogue and illustrate species that may be of interest to readers. There are about 260,000 species of plants, and many of them are attached to some mythology, superstition or folklore. For this volume, we have selected the more common ones, such as important food crops or generally available ornamental plants. We have prepared 83 entries, each comprehensively presented in informative sections, featuring more than 120 plants which readers are likely to encounter in daily life.

Each plant description will include the plant's scientific name, scientific and common family names, common name(s) in English, French, German, Hindi, Malay/Indonesian, Mandarin and Spanish, natural distribution, brief description of the plant, and the folklore, myths and other information pertaining to its auspiciousness. What one culture considers auspicious may be deemed inauspicious in another, and these varying perspectives have been included. Some of the more technical terms are explained in the Glossary, but, in general and whenever possible, simple yet accurate terminology has been used. Some gardening tips have been included for readers interested in cultivating some of these species in their gardens but most likely the auspicious ones!

Information on the plants was gathered from primary and secondary sources, such as interviews with people of different races and faiths and from various countries as well as from publications in print and on the Internet. The secondary sources in print have been credited in the References chapter.

It was a very interesting and enjoyable experience writing this book and we hope you will enjoy reading it as much as we did writing it.

Hugh T. W. Tan & Giam Xingli
August 2008

Introduction

What are plants?

Plants are those organisms that belong to the Plant Kingdom, one of kingdoms of life on Earth, the others being the viruses, archaea ('ancient bacteria'), bacteria, fungi, protists (algae, protozoans, slime and water moulds) and animals. Except for some parasitic or saprophytic species, plants are usually green in colour because they contain chlorophyll, which they use to photosynthesise to produce starch from sunlight, water and carbon dioxide. Plants are divided into the following extant groups: the mosses, liverworts and hornworts, club and spike mosses, horsetails, ferns, cycads, ginkgos, conifers, gnetophytes and flowering plants.

The plants in this book are mostly flowering plants (herbaceous to woody plants which possess water and food-conducting tissues and produce leaves, flowers, fruits and seeds), followed by some conifers (woody plants which possess water and food-conducting tissues and produce leaves, cones and seeds) and, least of all, the ferns (usually herbaceous plants which possess water- and food-conducting tissues and produce vegetative and spore-producing leaves but no seeds).

Sacred or magical plants; plants of mythology and superstition

Plants and humans are tightly intertwined, since plants are absolutely essential for food, providing numerous materials for building homes as well as fuel for cooking and keeping warm. Plants are used for medicines, as fodder for animals, and products and services too numerous to mention here. Plants also cater to the spiritual, religious and cultural aspects of humans.

In this book, we have defined auspicious plant in a broad sense to refer to any plant that is revered, considered sacred, favourable to grow for good fortune, health, prosperity or for protection from calamity, misfortune or evil spirits. Inauspicious plants are those that are considered unlucky, to invite calamity or ill health, and seen to be the abode of ghosts or evil spirits. Based on these definitions, the book covers plants which are auspicious and/or inauspicious.

How to Use this Book

Species of plants that are more commonly known and easily available to varied cultures of the world have been selected for description in this book and arranged, in alphabetical order, by their scientific names or plant grouping. Entries may also be for a plant group based on folklore and belief; in this book the plant groups are the Christmas tree, the four leaved clover, the heavily fruiting plants, and the prickly, spiny or thorny plants.

Scientific names are given more emphasis, as opposed to common names, as these provide a standard international reference to the identity of the plants. Common names, however, may be potentially confusing as the same plant may have many common names depending on the region it is found or cultivated. Each entry consists of the following:

FAST FACTS

Scientific Species Name, Scientific Hybrid Name or Scientific Name: This refers to the most preferred or up-to-date scientific name or botanical name of the plant species or hybrid. The scientific name of poinsettia, for example, consists of a generic name, *Euphorbia*, and a specific epithet, *pulcherrima*. A hybrid name has a multiplication sign (×) prefixing the generic name as in ×*Citrofortunella microcarpa* (lime-quat, an hybrid of two species from different genera, *Citrus* and *Fortunella*), or the specific epithet, *Ilex* ×*altaclerensis* (holly, a hybrid between two *Ilex* species). Synonym(s), a name which refers to the same species, is/are included too. Synonyms are encased within brackets and denoted by a preceding equals sign (=). For example, the combination, = *Curcuma domestica*, indicates *Curcuma domestica* is the synonym of *Curcuma longa* (turmeric). The etymology (derivation) of the scientific name is also included in brackets after the scientific name and its synonym(s), if applicable.

Common Species Name: Scientific names are in Latin, whereas the common or vernacular name of the plant species or hybrid is in the common native language name. Because of space constraints, only the common names of some of the world's major languages, where applicable, are given in this order: English, French, German, Hindi, Malay/Indonesian, Mandarin and Spanish. The Chinese language has a common written form, although the pronunciation varies with the dialect. We have opted for the Mandarin pronunciation since it is the official spoken form for the People's Republic of China and Taiwan. Mandarin names feature the Hanyu Pinyin form (the most common standard romanisation system for Mandarin), and the Chinese characters in simplified/traditional forms, e.g., the pineapple is *fèng lí* [凤梨/風梨].

Scientific Family Name: In scientific classification, species are grouped into genera (plural for genus) and the genera, in turn, are grouped into families. Each family has a scientific name and usually has the suffix -*ceae*, e.g., *Rutaceae* (orange family) and *Zingiberaceae* (ginger family). Some family names have other suffices such as -*neae* in the example of *Gramineae* (grass family), and -*tae*, as apparent in *Labiatae* (mint family). Synonym(s) is/are also included so that older literature or references to the plant are made more comprehensible. The classification and nomenclature for the families follow those of the Angiosperm Phylogeny Website (Stevens, 2001 onwards).

Common Family Name: This is the vernacular name of the family, usually named after one of the more famous species found in the family.

Natural Distribution: This refers to the plant's natural distribution or dispersion, as determined by experts through scientific research. Deliberate or accidental introductions by humans to a particular country do not count. For controversial cases, we have made a judgment call and selected that which we think is the most likely. Readers are most welcome to disagree!

MAIN FACTS

How to Grow: Plant care needs are provided here in this format.

Degree of exposure:

Requires full sunlight  Requires semi-shade 

Requires quarter-shade  Requires shade 

Water needs:

Requires much watering 

Requires moderate watering

Requires little watering 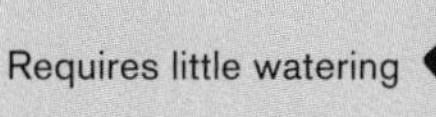

Aquatic plant 

USDA Plant Hardiness Zone(s):
These are 11 zones for growing plants of agriculture or the natural landscape (outdoor growth with adequate water) in the USA, Canada and Mexico, and which are primarily defined by their average annual minimum temperatures based on data of 1974–1986 for USA and Canada, and 1971–1984 for Mexico. Zones 2 to 10 are divided into two subzones and designated by the letters a and b, e.g., 2a, 2b. The smaller the number, the colder is the minimum temperature of its zone.

Zone 11 has a minimum temperature of 4.4°C (40°F), so has no frost and the warmest zone in this scheme. Also, the lower the letter of the alphabet, the colder it is, e.g., Zone 2a is colder than Zone 2b. The table below and on the next page shows the average annual minimum temperature ranges for the zones and subzones.

USDA Plant Hardiness Zone	Average Annual Minimum Temperature Range
1	<–50° F (<–45.56° C)
2a	–50 to <–45° F (–45.56 to <–42.78° C)
2b	–45 to <–40° F (–42.78 to <–40.00° C)
3a	–40 to <–35° F (–40.00 to <–37.22° C)
3b	–35 to <–30° F (–37.22 to <–34.44° C)
4a	–30 to <–25° F (–34.44 to <–31.67° C)
4b	–25 to <–20° F (–31.67 to <–28.89° C)
5a	–20 to <–15° F (–28.89 to <–26.11° C)
5b	–15 to <–10° F (–26.11 to <–23.33° C)
6a	–10 to <–5° F (–23.33 to <–20.56° C)
6b	–5 to <0° F (–20.56 to <–17.78° C)

MAIN FACTS

USDA Plant Hardiness Zone	Average Annual Minimum Temperature Range
7a	0 to <5° F (–17.78 to <–15.00° C)
7b	5 to <10° F (–15.00 to <–12.22° C)
8a	10 to <15° F (–12.22 to <–9.44° C)
8b	15 to <20° F (–9.44 to <–6.67° C)
9a	20 to <25° F (–6.67 to <–3.89° C)
9b	25 to <30° F (–3.89 to <–1.11° C)
10a	30 to <35° F (1.11 to <1.67° C)
10b	35 to <40° F (1.67 to <4.44° C)
11	40 to <45° F (4.44 to <7.22° C)

For more details, please see The United States National Arboretum USDA Plant Hardiness Zone Map (http://www.usna.usda.gov/Hardzone/index.html#intro). The zone system also applies to other regions such as Europe.

These data are indicated as such:
USDA x–y = USDA Plant Hardiness Zones x to y
USDA ≥x = USDA Plant Hardiness Zones x and warmer zones

Australian Plant Hardiness Zone(s):
These are seven zones for growing plants in the outdoors with adequate water in Australia, and which are primarily defined by their average annual minimum temperatures.

For more details, please see Plant Hardiness Zones for Australia by Iain Dawson (http://www.anbg.gov.au/hort.research/zones.html). The USDA and Australian hardiness zone **approximate** equivalence are as follows.

Australian Plant Hardiness Zone (average annual minimum temperature range)	USDA Plant Hardiness Zone
<–50° F (<–45.56° C)	7b, 8a
2 (–10 to <–5° C; 14 to <23° F)	8b, 9a
3 (–5 to <0° C; 23 to <32° F)	9b, 10a
4 (0 to <5° C; 32 to <41° F)	10b, 11
5 (5 to <10° C; 41 to <50° F)	11
6 (10 to <15° C; 50 to <59° F)	None available
7 (15 to <20° C; 59 to <68° F)	None available

This is indicated as such:
AUS p–q = Australian Plant Hardiness Zones p to q
AUS ≥p = Australian Plant Hardiness Zones p and warmer zones

MAIN FACTS

Drawbacks: Also mentioned are the disadvantages of growing the plant, such as diseases, pests and other possible inconveniences.

Plant Description: The key features and characteristics of the plant are mentioned here. Technical terms are explained in the **Glossary**.

Folklore, Beliefs, Uses and Other Interesting Facts:
Various religious, superstitious and mythical associations tied to each plant entry are listed alongside some major economic uses or other interesting information. Feng shui beliefs are also indicated. Homonymies (the homonym being the same common name referring to another species) as well as close plant relatives are highlighted. For brevity and reader convenience, the following symbols apply:

 Auspicious plant

 Inauspicious plant

 Evil spirit plant

 Spirit plant

 Plant for protection against evil or harm

 Chinese (Lunar) New Year plant

 Christmas plant

 Buddhist belief

 Christian belief

 Hindu belief

 Judaism belief

 Islamic belief

The plants in this book have these features.

Key to Plant Features: A = aquatic herb; C = climber; H = herb; S = shrub; T = tree;
1 = yes; 0 = no; NA = not applicable

	Scientific Name	Temperate	Subtropical	Tropical	Plant Habit	Auspicious	Inauspicious	Evil Spirit Plant	Spirit Plant	Protective	Chinese New Year	Christmas	Buddhist Belief	Christian Belief	Hindu Belief	Judaism Belief	Muslim Belief
1	*Adenium obesum*	0	1	1	S	1	0	0	0	0	0	0	0	0	0	0	0
2	*Aegle marmelos*	0	1	1	T	1	1	0	0	1	0	0	0	0	1	0	0
3	*Alstonia scholaris*	0	0	1	T	0	1	1	0	0	0	0	0	0	1	0	0
4	*Ananas comosus*	0	1	1	H	1	1	0	0	1	1	0	0	0	0	0	0
5	*Azadirachta indica*	0	1	1	T	1	0	0	0	1	0	0	0	0	1	0	0
6	Bambuseae	1	1	1	ST	1	0	0	0	1	0	0	0	0	0	0	0
7	*Betula* species	1	0	0	ST	1	0	0	0	1	0	0	0	0	0	0	0
8	*Camellia japonica*	1	1	0	ST	1	0	0	0	0	1	0	0	0	0	0	0
9	*Cananga odorata*	0	0	1	T	0	1	1	1	0	0	0	0	0	0	0	0
10	*Canavalia gladiata*	0	0	1	CS	1	0	0	0	0	1	0	0	0	0	0	0
11	*Cardamine pratensis*	1	0	0	H	0	1	0	0	1	0	0	0	0	0	0	0
12	*Castanospermum australe*	0	1	1	T	1	0	0	0	0	1	0	0	0	0	0	0
13	*Ceiba pentandra*	0	1	1	T	0	1	1	0	0	0	0	0	0	0	0	0
14	*Cercis siliquastrum*	1	0	0	ST	0	1	0	0	0	0	0	0	1	0	0	0
15	Christmas tree	1	0	0	T	1	0	0	0	0	0	1	0	0	0	0	0
16	*Cibotium barometz*	0	1	1	H	1	0	0	0	0	1	0	0	0	0	0	0
17	x*Citrofortunella microcarpa*	0	1	1	ST	1	0	0	0	0	1	0	0	0	0	0	0
18	*Citrus medica* var. *sarcodactylis*	0	1	1	ST	1	0	0	0	1	1	0	1	0	0	0	0
19	*Citrus reticulata*	0	1	1	ST	1	0	0	0	0	1	0	0	0	0	0	0
20	*Cocos nucifera*	0	1	1	T	1	1	0	0	1	0	0	0	0	1	0	0

	Scientific Name	Temperate	Subtropical	Tropical	Plant Habit	Auspicious	Inauspicious	Evil Spirit Plant	Spirit Plant	Protective	Chinese New Year	Christmas	Buddhist Belief	Christian Belief	Hindu Belief	Judaism Belief	Muslim Belief
21	*Conium maculatum*	1	0	0	H	0	0	1	0	0	0	0	0	0	0	0	0
22	*Cordyline fruticosa*	0	1	1	S	1	0	0	0	1	0	0	0	0	0	0	0
23	*Crassula ovata*	0	1	1	H	1	0	0	0	0	1	0	0	0	0	0	0
24	*Crataegus* species	1	0	0	ST	1	1	0	1	1	0	0	0	1	0	0	0
25	*Curcuma longa*	0	1	1	H	1	0	0	0	0	0	0	0	0	1	0	0
26	*Dendranthema xgrandiflora*	1	1	0	H	1	1	0	0	0	1	0	0	0	0	0	0
27	*Dianthus caryophyllus*	1	1	0	H	1	1	0	0	0	0	0	0	0	0	0	0
28	*Dracaena fragrans*	0	1	1	ST	1	0	0	0	1	0	0	0	0	0	0	0
29	*Dracaena sanderana* or *Dracaena sanderiana*	0	1	1	HS	1	0	0	0	0	1	0	0	0	0	0	0
30	*Epiphyllum oxypetalum*	0	1	1	S	1	1	1	1	0	0	0	0	0	0	0	0
31	*Euphorbia pulcherrima*	0	1	1	ST	1	0	0	0	0	1	1	0	0	0	0	0
32	*Ficus benjamina*	0	1	1	T	1	1	1	1	0	0	0	0	0	0	0	0
33	*Ficus carica*	1	0	0	ST	0	0	0	0	0	0	0	0	1	0	1	1
34	*Ficus microcarpa*	0	1	1	T	1	1	1	1	0	0	0	0	0	0	0	0
35	*Ficus religiosa*	0	1	1	T	1	0	0	0	0	0	0	1	0	1	0	0
36	*Fortunella japonica* or *Fortunella margarita*	1	1	1	ST	1	0	0	0	0	1	0	0	0	0	0	0
37	Four leaved clover	1	1	1	H	1	0	0	0	1	0	0	0	1	0	0	0
38	*Fuchsia* species and hybrids	1	1	0	ST	1	0	0	0	0	1	0	0	0	0	0	0
39	*Gladiolus* species and hybrids	1	1	0	H	1	1	0	0	1	1	0	0	0	0	0	0
40	Heavily fruiting plants	1	1	1	CH ST	1	0	0	0	0	1	0	0	0	0	0	0

The plants in this book have these features.

Key to Plant Features: A = aquatic herb; C = climber; H = herb; S = shrub; T = tree;
1 = yes; 0 = no; NA = not applicable

	Scientific Name	Temperate	Subtropical	Tropical	Plant Habit	Auspicious	Inauspicious	Evil Spirit Plant	Spirit Plant	Protective	Chinese New Year	Christmas	Buddhist Belief	Christian Belief	Hindu Belief	Judaism Belief	Muslim Belief
41	*Hedera helix*	1	1	0	C	1	1	0	0	1	0	1	0	0	0	0	0
42	*Hordeum vulgare*	1	1	0	H	0	0	0	0	0	0	0	0	1	0	1	0
43	*Hyacinthus orientalis*	1	0	0	H	1	0	0	0	0	1	1	0	0	0	0	0
44	*Hypericum* species	1	0	0	H	1	0	0	0	1	0	0	0	1	0	0	0
45	*Ilex* species and hybrids	1	0	0	ST	1	0	0	0	1	0	1	0	0	0	0	0
46	*Jasminum sambac*	0	1	1	CS	0	0	0	1	1	0	0	0	0	1	0	0
47	*Juniperus* species and hybrids	1	1	1	ST	1	1	0	0	1	0	0	0	1	0	0	0
48	*Lageneria siceraria*	0	1	1	C	1	0	0	0	1	1	0	0	0	0	0	0
49	*Laurus nobilis*	1	1	1	ST	0	0	0	0	1	0	0	0	0	0	0	0
50	*Lilium* species and hybrids	1	1	0	H	1	1	0	0	1	0	0	0	1	0	0	0
51	*Mangifera indica*	0	1	1	T	1	0	0	0	0	0	0	1	0	1	0	0
52	*Michelia xalba*	0	1	1	T	0	0	1	1	0	0	0	0	0	1	0	0
53	*Michelia champaca*	0	1	1	T	0	0	1	1	1	0	0	0	0	1	0	0
54	*Musa* species and hybrids	0	1	1	H	1	0	1	1	0	0	0	0	0	1	0	0
55	*Nelumbo nucifera*	1	1	1	A	1	0	0	0	0	0	0	1	0	1	0	0
56	*Nyctanthes arbor-tristis*	0	1	1	ST	0	0	0	0	0	0	0	0	0	1	0	0
57	*Ocimum tenuiflorum*	0	1	1	H	1	0	0	0	1	0	0	0	0	1	0	0
58	*Olea europaea*	1	1	1	ST	1	0	0	0	0	0	0	0	1	0	1	1
59	*Oryza sativa*	1	1	1	H	1	0	0	0	0	1	0	0	0	0	0	0
60	*Pachira aquatica*	0	1	1	T	1	0	0	0	0	1	0	0	0	0	0	0
61	*Paeonia* cultivars	1	0	0	HS	1	1	0	0	1	1	0	0	0	0	0	0

	Scientific Name	Temperate	Subtropical	Tropical	Plant Habit	Auspicious	Inauspicious	Evil Spirit Plant	Spirit Plant	Protective	Chinese New Year	Christmas	Buddhist Belief	Christian Belief	Hindu Belief	Judaism Belief	Muslim Belief
62	*Phoenix dactylifera*	1	1	0	T	1	0	0	0	1	0	0	0	1	0	1	0
63	*Pinus* species	1	1	1	ST	1	1	0	0	1	0	0	0	0	0	0	0
64	*Platycladus orientalis*	1	1	1	T	1	0	0	0	1	0	0	0	0	0	0	0
65	Prickly, spiny or thorny plants	1	1	1	CH ST	0	1	0	0	1	0	0	0	1	0	0	0
66	*Prunus mume* and *Prunus salicina*	1	0	0	T	1	0	0	0	0	1	0	0	0	0	0	0
67	*Prunus persica* cultivars	1	0	0	T	1	0	0	0	1	1	0	0	0	0	0	0
68	*Prunus serrulata* and *Prunus xyedoensis*	1	1	0	T	1	0	0	0	0	0	0	0	0	0	0	0
69	*Punica granatum*	1	1	1	S	1	1	0	0	1	0	0	1	1	0	1	0
70	*Rosa* species and hybrids	1	1	1	CS	0	1	0	0	1	0	0	0	1	0	0	0
71	*Rosmarinus officinalis*	1	1	0	S	1	1	0	0	1	0	0	0	0	0	0	0
72	*Saccharum officinarum*	0	1	1	H	1	1	0	0	0	0	0	0	0	1	0	0
73	*Salix caprea*	1	1	1	ST	1	0	0	0	0	1	0	0	1	0	0	0
74	*Salix* species and hybrids	1	1	1	ST	0	1	0	0	1	0	0	0	1	0	0	0
75	*Saraca asoca*	0	1	1	T	1	0	0	0	1	0	0	1	0	1	0	0
76	*Solanum mammosum*	0	1	1	S	1	0	0	0	0	1	0	0	0	0	0	0
77	*Thuja occidentalis*	1	0	0	T	0	0	0	0	1	0	0	0	0	0	0	0
78	*Thysanolaena latifolia*	1	1	1	H	1	0	0	0	0	0	0	1	0	0	0	0
79	*Vicia faba*	1	1	0	H	0	1	0	0	0	0	0	0	0	0	0	0
80	*Viscum album*	1	0	0	S	0	0	0	0	0	0	1	0	0	0	0	0
81	*Vitis vinifera*	1	1	1	C	0	0	0	0	0	0	0	0	1	0	1	0
82	*Zamioculcas zamiifolia*	0	0	1	H	1	0	0	0	0	1	0	0	0	0	0	0
83	*Zantedeschia aethiopica*	1	1	0	H	0	1	0	0	0	0	0	0	0	0	0	0
	Totals	46	64	55	NA	62	28	10	9	35	26	6	7	16	16	6	2

The Plants

Adenium obesum desert rose • impala lily

Scientific Species Name:
Adenium obesum
(Greek *adenos*, gland; Latin *obesus*, fat, referring to the swollen basal stems and roots of this plant)

Common Species Name:
desert rose, impala lily (English); *fù guì huā* [富贵花/富貴花] (Mandarin)

Scientific Family Name:
Apocynaceae

Common Family Name:
periwinkle family

Natural Distribution:
East Africa to Southern Arabia

USDA ≥11 (AUS ≥5)

How to Grow: Well-drained soil; water only when necessary, as does not tolerate damp conditions because it is a desert plant, and keep drier during winter; leaf drop during winter in temperate climates but evergreen in tropical areas; propagation by stem cuttings, seed or scion grafting.

Drawbacks: Roots prone to rotting when over-watered; generally pest-free because of the poisonous latex, and trouble-free if not over-watered.

Plant Description:

- A medium-sized succulent shrub which grows up to about 2 m (6.6 ft) tall. The swollen basal stem merges into the equally swollen roots, particularly in older plants.
- The alternate, shortly stalked leaves grow up to 15 cm (5.9 in) long and the leaf blades are obovate (or egg-shaped, with the wider end outward), thinly leathery, shiny dark green above, light green below.
- Flowers arranged in clusters at the branch tips. The 5 petals are joined into a trumpet like structure, with the free ends radiating outwards, 5–8 cm (2–3.1 in) across, red to purple with a white eye, or solid white.
- Each flower produces a pair of horn-like fruits which split open when ripe to release the hairy-winged seeds.

Adenium obesum plant, which is a scion graft

 Photos: Hugh Tan Tiang Wah

Plant with swollen stem base and roots, in a sun rockery

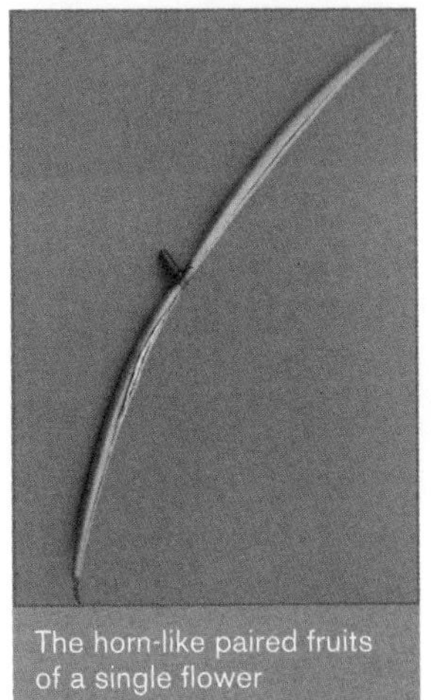

The horn-like paired fruits of a single flower

Splitting fruit and seed

Folklore, Beliefs, Uses and Other Interesting Facts

- The Mandarin name for the desert rose, *fù guì huā*, signifies a life of fortune and comfort, the 'flower of wealth'. This plant is believed by the Chinese to bring such a life to its owner, hence it is an auspicious plant to cultivate in pots or dragon jars. The cultivars with pink or red flowers are more valued than those with white ones because it is believed that red or brightly-coloured flowers are like life-blood, while white flowers are used for funerals and connected to death and sorrow. The swollen, stout stem base and roots represent fertility and abundance, and therefore the plants with the most swollen parts are always held in the highest regard.
- According to feng shui beliefs, red- or pink-flowered plants are auspicious for growing at the front, south, southwest and northeast of the garden; bad for the west and northwest; and neutral for the other sectors of the garden. While the desert rose can be cultivated as a bonsai, feng shui belief is to avoid such plants because they signify stunted growth.
- The plant is a source of poison, which is used to coat the tips of arrows or for ordeal rituals to determine guilt or innocence.
- The plant can also be grown as an ornamental.

Aegle marmelos bael fruit • Indian quince

Scientific Species Name:
Aegle marmelos
(Greek *aigle*, the light of the sun; Portuguese *marmelo*, quince)

Common Species Name:
bael, bael fruit, bael tree, belfruit tree, Bengal quince, golden apple, holy fruit, Indian bael fruit, Indian quince, stone apple, wood apple (English); *bel Indien, oranger du Malabar* (French); *Belbaum, Bengalische Quitte, Indische Quitte* (German); *bel, bael, beli, belgiri, sirphal* (Hindi); *bila, bilak* (Malay); *mù jié* [木桔] (Mandarin); *bela, milva* (Spanish). Occasionally called elephant apple, which also refers to *Dillenia indica* or *Feronia limonia*

Scientific Family Name:
Rutaceae

Common Family Name:
citrus family

Natural Distribution:
Pakistan, Central and Southern India, Nepal, Bangladesh, Myanmar, Cambodia, Laos and Vietnam

USDA ≥9 (AUS ≥3)

How to Grow: Seasonal tropical or subtropical plant; rich well-drained soil of pH 5–8; drought-tolerant, requiring a long dry season to fruit; minimal fertiliser required; propagation by seed, bud grafting, shield budding, air-layering or root cuttings.

Drawbacks: Untidy tree, so needs pruning to keep good shape; thorny stems and branches; requires a long dry season to fruit, so not entirely suitable for the wet tropics; relatively free from pests and diseases but may have sooty moulds on leaves (apply white summer oil to eradicate).

Plant Description:

- Slow-growing, deciduous tree growing to 15 m (49.2 ft) tall.
- It has thick, soft, flaking bark on the trunk. A clear gummy sap that hardens flows from wounded branches. Branches and the young stem are sometimes thorny and turn light brown when older.
- Leaves are alternate, composed of 3–5 leaflets, pale green when young, and aromatic.
- Bears small, greenish-white, fragrant, bisexual flowers that are borne in 4–7-flowered clusters which arise from the leaf axils. Each flower has 4 fleshy petals and numerous stamens.
- The round, yellow-green fruit is 5–7 cm (2.0–2.8 in) across, embellished with small dots (oil glands). The sweet-tasting, yellow-orange coloured pulp is thickly mucilaginous and is embedded with numerous hair-covered seeds. Though there are thin-rind cultivated varieties, many plants have very hard-shelled fruits that can only be cracked open with a hammer or sawed through.

The fruit and leafy branches of *Aegle marmelos*

 Photos: (above) Angie Ng-Chua; (opposite) Hugh Tan Tiang Wah

Newly developed and mature leaves have three leaflets signifying the three functions of Lord Siva

A thorny branch

Folklore, Beliefs, Uses and Other Interesting Facts

- The bael fruit is an important sacred tree in Hindu folklore and, therefore, often cultivated in temple gardens in India. See Gupta (1971) for details of Hindu folklore pertaining to the bael fruit.
- The three leaflets of the leaf represent the trident, which is the symbol of Lord Siva of Hindu mythology. The leaflets signify the three functions of Siva—creation, preservation and destruction—and his three eyes.
- Aromatic bael fruit leaves are used as offerings in a compulsory ritual in worship of Lord Siva. Its medicinal properties also contribute to its status as a sacred plant; many parts of this tree are used for curing a variety of ailments.
- It can be used medicinally as an antibiotic, antiemetic, digestive aid, diuretic, mild laxative, sedative and tonic; in treating ailments including asthma, bronchitis, constipation, diarrhoea, dysentery, fever, heart palpitation, haemorrhoids, inflammation, jaundice, leucoderma, malaria, ophthalmia, rhinitis, snakebite.
- Indian folklore dictates that the tree is planted on the north side of the house.
- Feng shui belief, however, recommends placing this plant away from the house because it possesses long, sharp thorns (signifying poison arrows that send out hostile, inauspicious energy), but it can be planted near the gate as a sentinel.
- The bael fruit pulp is eaten or made into refreshing drinks.

Alstonia scholaris devil tree • blackboard tree

Scientific Species Name:
Alstonia scholaris
(*Alstonia*, named after Dr. Charles Alston (1685–1760), the Scottish botanist and professor of Botany, Edinburgh University; Latin *scholaris*, belonging to a school–referring to the writing exercises children in 19th-century Southeast Asia did on wooden boards of this species)

Common Species Name:
blackboard tree, devil tree, dita bark, Indian devil tree, Indian pulai, milky pine, milkwood pine, pulai, white cheesewood (English); *pulai* (Malay); *táng jiāo shù* [糖胶树/糖膠樹] (Mandarin); *saptaparna* (Sanskrit)

Scientific Family Name:
Apocynaceae

Common Family Name:
periwinkle family

Natural Distribution:
India, Sri Lanka, Southeast Asia, Tropical Australia, Solomon Islands

USDA ≥11 (AUS ≥5)

How to Grow: Seasonal tropics; most soils; tolerant of wet conditions; propagate by seed.

Drawbacks: Generally pest and disease-free.

Plant Description:
- A tree that grows to about 35 m (114.8 ft) tall, with a pagoda-like crown of tiered branches.
- The stalked leaves are in whorls of 3–8 leaves, which grow up to 20 cm (7.9 in) long. The leaf blade is oblanceolate and leathery.
- Its strongly fragrant flowers are pale greenish-yellow and they are arranged in tight clusters growing from leaf axils.
- Each flower usually produces 2 elongated, pod-like follicles, which split when ripe to release hairy winged seeds which are wind-dispersed.

Photos: (opposite top and middle) David Lee; (opposite bottom) Hugh Tan Tiang Wah

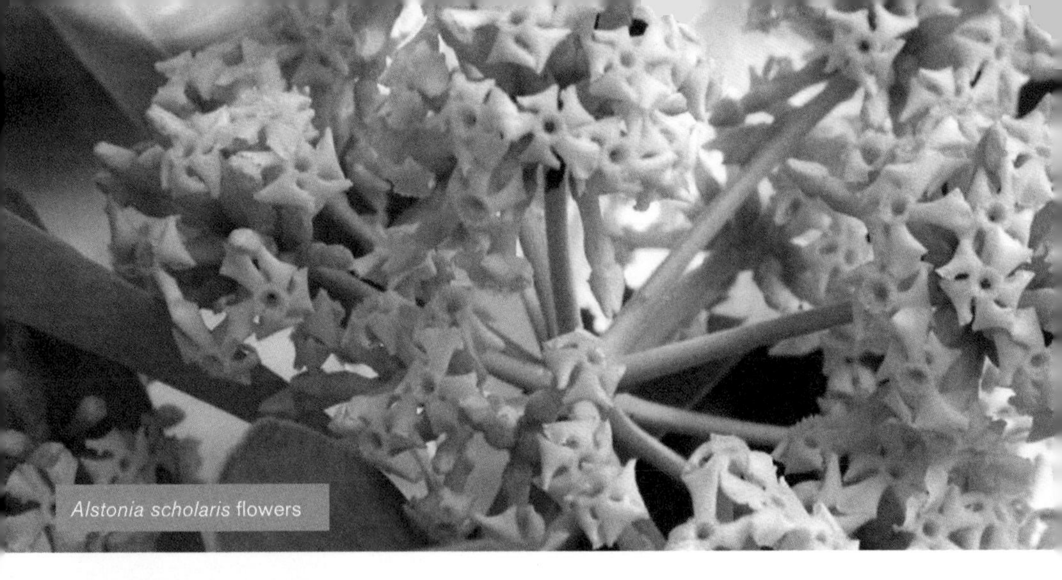
Alstonia scholaris flowers

Ripe (brown) and unripe (green) fruits

Tree with its tiered branches

Folklore, Beliefs, Uses and Other Interesting Facts

- A Hindu belief is that *Alstonia scholaris* is the tree of the devil.
- In 19th-century Southeast Asia, wood from the tree was used to make wooden slates for school children to write on, hence the specific epithet *scholaris* in the scientific name. Students wrote with a pen made of *Arenga* (palm) wood on the pipe-clay coated wood of the blackboard tree. The wood was also used to make coffins.
- This plant has numerous medicinal uses (including Indian Ayurvedic medicine and Malay folk medicine), owing largely to the numerous alkaloids in the poisonous latex. It can be used as a vermifuge, an alleged anti-malarial, an antipyretic, and to treat skin disorders and urticaria (hives).
- This species is very similar to *pulai* (*Alstonia angustiloba*) except that the *pulai* has white rather than greenish yellow flowers, smooth rather than minutely pimply seeds and a rather odourless flower, plus a few other distinguishing features.

Ananas comosus pineapple

Scientific Species Name:
Ananas comosus
(*Ananas*, from *nanas*, the South American indigenous (Tupi) name for the pineapple; Latin *comosus*, tufted, referring to the leafy 'crown' topping the pineapple fruit)

Common Species Name:
pineapple (English); *ananas commun* (French); *Ananas* (German), *anannaasa* (Hindi); *nanas* (Malay); *bō luó* [波萝], *fèng lí* [风梨/風梨], *huáng lí* [黄梨], *wáng lí* [王梨] (Mandarin); *ananá*, *piña*, *piña de américa*, *piña tropical* (Spanish)

Scientific Family Name:
Bromeliaceae

Common Family Name:
bromeliad family

Natural Distribution:
All current members are cultigens with no wild ancestral forms present. Wild ancestral forms were native to tropical South America.

USDA ≥11 (AUS ≥5)

How to Grow: A tropical herb sensitive to frost; fairly drought tolerant; well-drained, high organic content, sandy loam with pH 4.5–6.5; propagate by suckers from the plant base or 'crowns' at the fruit tops, using seeds only for breeding, after hand-pollinating the flowers.

Drawbacks: Spiny plant so handle with care; animal pests include nematodes, mealy bugs, pineapple mite, sap beetle, butterfly or moth caterpillars, cutworms, rats, crows, mice, rabbits; diseases include top, root and basal rots, and wilt by fungi, and yellow spot virus.

Plant Description:

- Perennial herb growing to 1.5 m (4.9 ft) tall.
- Sword-like spiral leaves with spiny margins are arranged in a basal rosette.
- Each sharp-tipped leaf has a long sword-like shape and is rigid with spiny margins. Each leaf can grow up to 80 cm (31.5 in) long.
- The flowers are arranged in an erect, cylindrical inflorescence of up to 30 cm (11.8 in) long at the stem tip.
- The pineapple fruit is a fleshy composite fruit made up from the ovaries of the flowers, with juicy, off-white to golden-yellow flesh inside.
- In the wild, hummingbirds are the pollinators, and small, hard seeds develop after the flowers are pollinated and fertilisation occurs. In cultivation in the Old World tropics, where there are no hummingbirds, the flowers do not get pollinated and no seeds form.

Ananas comosus fruit

Photos: (above and opposite) Hugh Tan Tiang Wah

Fruit on display at a greengrocer's shop

Mutant fruit with numerous suckers sold at Chinese New Year as an auspicious ornament

Folklore, Beliefs, Uses and Other Interesting Facts

- The name in Hokkien (the dialect of Fujian Province, China) for the pineapple is *ong lai*, which is a phonetic pun for 'coming prosperity'. Hence the fruit and its images on crystalware (this is lead crystal moulded into the shape of a pineapple fruit) and plastic ornaments is considered auspicious in the Chinese, especially Hokkien, home. The golden-yellow of the ripe fruit also represents auspiciousness and royalty. In Peninsular Malaysia and Singapore, mutant pineapple fruits with more leafy shoots are also used as auspicious decoration during the Chinese New Year.
- A popular Chinese old wives' tale warns pregnant women not to consume pineapple fruit or juice for fear of a miscarriage. This is not completely without reason as large doses of pineapple juice have been reported to cause uterine contractions.
- Based on feng shui beliefs, the pineapple plant, notwithstanding the auspicious fruit, with its spiny leaves, is considered inauspicious because the spines represent 'poison arrows'. One can plant it as a sentinel plant near the gate but should never grow it indoors because the spines' poison arrows are believed to emit hostile, inauspicious energy.
- The pineapple is cultivated for its delicious fruit from numerous cultivars. The fruit is eaten fresh, canned, frozen, dried, or made into juices and syrups. The residue, after processing the fruit, is used as livestock feed.
- The meat tenderiser, bromelain, is extracted from the pineapple fruit.
- Strong, white, silky fibre is extracted from pineapple leaves for making thread and textiles for clothing.

Azadirachta indica neem • nimtree

Scientific Species Name:
Azadirachta indica
(= *Melia indica*, *Melia azadirachta*)
(Latin *azadirachta* is derived from the Persian *azad dirakht*, noble tree; Latin *indica*, of India)

Common Species Name:
bead tree, Burmese neem tree, Chinaberry, Indian cedar, Indian lilac, margosa, margosa tree, neem, neem tree, nimtree (English); *azadirachta de l'Inde*, *lilas des Indes*, *neem des Indes*, *nim des Indes*, *margosier*, *margosier de Birmanie*, *margosier du Bangladesh* (French); *Niembaum*, *Nimbaum* (German); *balnimb*, *neem*, *nim*, *nind* (Hindi); *intaran*, *mambu*, *mind*, *sadu* (Malay); *liàn shù* [楝树/楝樹], *kǔ liàn* [苦楝], *liàn zào zǐ* [楝枣子/楝棗子], *yìn dù liàn shù* [印度楝树/印度楝樹] (Mandarin)

Scientific Family Name:
Meliaceae

Common Family Name:
mahogany family

Natural Distribution:
Probably Myanmar, but it has been widely cultivated and naturalised in the tropics, even becoming an invasive in Fiji, Mauritius and others.

USDA ≥11 (AUS ≥5)

How to Grow: Cold and frost intolerant but tolerant of high temperatures; tolerant of a 7–8-month dry season; slightly acid to slightly alkaline soil (pH 5.0–8.0); propagated by seed (usually), stem cuttings or root suckers.

Drawbacks: Generally disease and pest-free; insect pests include scale, leaf-cutting ants, tortricid and pyralid moth larvae, bug attack which may be overcome with appropriate insecticide application; diseases include fungal ones such as root rot (wet ground), blight (stems and twigs) and leaf spot, and a bacterial blight (leaves); intolerant of waterlogged ground.

Plant Description:

- An evergreen tree that grows to 30 m (98.4 ft) tall and 0.8 m (31.5 in) trunk diameter. It sheds leaves during prolonged dry periods.
- The stalked leaves are alternate and pinnate. The leaflets are almost stalkless, opposite and the blades are somewhat sickle-shaped, with toothed margins.
- The sweetly scented, small, bisexual flowers are produced on elongated shoots at the leaf axils. The 5 sepals are joined into a small cup, the 5 petals are white, the 10 stamens are joined into a tube and the pistil consists of an ovary and 1 style.
- The fruit is olive-like, round or oblong, about 1–2 cm (0.4–0.8 in) long, light yellow when ripe.
- Each fruit has a single seed.

Azadirachta indica fruits

Photos: (above and opposite) Hugh Tan Tiang Wah

Azadirachta indica tree

Flowering and leafy branches. Note the young reddish green leaves.

Top and bottom views of leaflets

Folklore, Beliefs, Uses and Other Interesting Facts

- The nimtree can grow on poor soils and is one of the world's most useful trees. It can be used for fuel, for shade (50,000 trees have been planted to shade the annual camp of 2 million Muslim Haj pilgrims in the plains of Saudi Arabia), and as an ornamental tree for streets, parks and gardens. Its flowers and leaves can be used as fodder, its timber a substitute for mahogany, and its seed-oils in soaps, toothpaste, lotions and insecticide (azadirachtin at 10 parts per million is lethal to most moth and butterfly larvae). It is also used against skin diseases (scabies, chicken pox) and in antimalarial, postcoital contraceptive and arthritis treatment. Nim is often used in Indian cuisine to add a refreshing flavour to dishes.
- Because of its numerous uses, nim is thought of as a divine tree by Indians, and planted in temples and homes.
- In India, the nimtree is regarded as a symbol of the female, whereas the male is represented by the bodhi tree (*Ficus religiosa*). In villages, the nim and bodhi trees are grown together, surrounded by a platform. Intertwined or coiled snake stones symbolising fertility are placed on the platform. The sexes represented by the trees are reversed in Punjab and Rajasthan where the nimtree is the male instead.
- A Hindu belief is that nim leaves placed inside pillowcases will protect weak-hearted and young children from spirits that might disturb them during their sleep.
- Based on feng shui principles, the nimtree with its open, airy and bright green crown is considered auspicious. It has white flowers and so is good for the west, northwest and north zones of the garden but bad for the east and southeast and neutral for the rest.

Bambuseae bamboo

Scientific Species Name:
Bambuseae
Tribe Bambuseae of subfamily Bambusoideae of the family Poaceae (After *Bambusa*, a genus in the Bambuseae; *Bambusa*, the Latinised version of the Malay *bambu*, bamboo)

Common Species Name:
bamboo (English); *bambou* (French); *Bambus* (German); *baans, baansa* (Hindi); *aur, bambu, buluh* (older spelling, *buloh*) (Malay); *zhú* [竹] (Mandarin); *bambú* (Spanish)

Scientific Family Name:
Poaceae or **Gramineae**

Common Family Name:
grass family

Natural Distribution:
Tropical to temperate, but mostly in the tropical to subtropical areas of Central and South America, Africa and Asia

How to Grow: As bamboos are from about 50 genera and a few hundred species, it is difficult to provide general information on their horticultural needs; readers are requested to look up the needs of individual species from other sources. Most species are generally hardy and prefer open conditions, damp and warm conditions but not soggy ground; propagation is usually by stem cuttings or splitting the clumps at the rhizome or by seed.

Drawbacks: Cells of the leaves often possess silica cells, which make them difficult to break down so leaf litter is a problem. These cells can scratch the paintwork of cars hence it is not a plant for car parks; some species flower gregariously after some decades with mass deaths, with new plants arising from the set seed or from remaining rhizomes.

Plant Description:

- Bamboos are woody, perennial evergreens of the grass family.
- Some characteristics of bamboos place them apart from the typical grasses: a woody (versus non-woody or herbaceous), hollow stem (known as a culm) and the existence of a strongly branching habit (versus little or no branching).
- Like many grasses, the culm develops from an underground horizontal rhizome.
- The culms can be erect, slightly pendulous or climbing. The nodes are visible and separated by hollow internodes.
- The ultimate branches usually each bear about 8–12 foliage leaves. These leaves are stalked, unlike the stalkless modified leaves. The culm sheaths or culm leaves, which develop lower down the culm, serve to enclose and protect the tender lower part of the internode (the part of the culm between two consecutive joints) of the culm while it is tender and growing.
- Bamboos have a compound inflorescence composed of florets (small flowers) in spikelets (small spikes, or elongated inflorescences with stalkless flowers) or pseudospikelets (inflorescences resembling spikelets).
- The fruit is a grain (caryopsis) comprising a fruit wall fused to the seed coat of one seed.

Dumpling wrapped in bamboo leaves

 Photos: (above and opposite) Hugh Tan Tiang Wah

Stems and leaves of the yellow-stemmed form of *Schizostachyum brachycladum*

Bambusa vulgaris 'Vittata' stems

Stems of *Bambusa vulgaris* 'Wamin'

Folklore, Beliefs, Uses and Other Interesting Facts

- The Chinese regard the bamboo plant as a symbol of longevity (like the pine), durability, endurance and hardiness because the culms of the bamboo are strong enough to withstand the weight of snow, flexible enough to withstand the force of typhoons and hurricanes, and resilient enough to withstand the changing of seasons (always green and flourishing). Its ability to remain erect and alive all year round epitomises the never-say-die attitude the Chinese strive to achieve.

- The Chinese associate bamboos with being disciplined and upright, being one of the 'Four Gentlemen of Flowers', representing summer. The other three 'gentlemen' are the plum (winter), the orchid (spring) and the chrysanthemum (autumn).

- Bamboos with specific morphologies are linked to different beliefs. For instance, the internodes on the culms of *Bambusa vulgaris* 'Wamin' (Buddha's belly bamboo) are basally inflated, making the stem look like a string of pearls. The shape of the internode is likened to the belly of the Laughing Buddha and therefore believed to bless its owner by virtue of morphological similarity with the holy being.

- Based on feng shui beliefs, bamboo plants are powerful symbols of good fortune, and a clump can be used to block off or deflect *shā qì* or 'killing breath' to a home from large construction features, such as a building, road or overpass, or the 'poison arrows' of tall narrow structures such as lamp posts, signboards and fences. Plant bamboo to the left of the house (when facing outwards) to signify the green dragon which brings abundance and prosperity; at the backyard for solid support (especially for those doing business); near the front to attract auspicious energy or *zhēn qì*; or anywhere else to represent longevity and good health. Grow a big clump in the east of the garden to energise the wood element, in particular. Avoid *bonsai* bamboo which symbolises stunted growth. Dead or old growth should always be removed so that the bamboo is a symbol of flourishing growth. Bamboo wind chimes or flutes hung in the home are thought to fend off evil spirits and bad *qì*. For prosperity in a shop, hang a pair of bamboo stems over the shop's entrance or on a wall facing the entrance; the stems should be tied with red ribbon (to energise them) and tilted toward each other.

Betula species birch

Scientific Species Name:
***Betula* species**
(Latin *betula*, birch)

Common Species Name:
birch (English); *bouleau* (French); *Birke* (German); *bhoj*, *bhurj*, *pattra* (Hindi); *birch* (Malay); *huà mù* [桦木/樺木] (Mandarin); *abedul* (Spanish)

Scientific Family Name:
Betulaceae

Common Family Name:
birch family

Natural Distribution:
Temperate and boreal northern hemisphere

USDA 2–9 (AUS 1–3)

How to Grow: There are about 35 birch species so these notes may not apply and do check for individual species in other references; fast growing trees; generally frost-tolerant; moist, acidic soils; water during dry weather; generally not drought tolerant; tolerates periodic flooding; may have low to no tolerance to salt aerosol; propagate by seed, semi-woody or woody stem cuttings or bud grafting.

Drawbacks: May produce a lot of leaf litter; roots may damage foundations and other structures; yellowing of leaves may occur with alkaline soils; may have leaf spot, canker disease or branch dieback (prune to increase vigour); may be susceptible to bronze birch borers in warmer areas.

Plant Description:

- Deciduous shrubs or trees that grow to 30 m (98.4 ft) tall, with one to many trunks.
- The trunk bark is often peeling, smooth and chalky white to dark brown.
- The alternate, simple, stalked leaves generally have leaf blades with a toothed, scallop-edged to shallowly lobed margin.
- Each flower is either male or female but both types of flowers are found on one plant, each type arranged in a unisexual flowering shoot (catkin). The male catkin is slender and drooping, single or clustered at the branch tip; the usually solitary female catkin is shorter and stiffer and develops below the male catkins.
- Fruits are small, winged nutlets (samaras) borne on the conifer cone-like female catkin.

Betula papyrifera fruits

Photos: (above) Steve Hurst @ USDA-NRCS PLANTS Database; (opposite) USDA-NRCS PLANTS Database

Leafy branch of *Betula papyrifera*

Peeling bark of *Betula papyrifera*

Betula papyrifera tree

Folklore, Beliefs, Uses and Other Interesting Facts

- In early Celtic mythology, the birch was said to represent purification and renewal because of its association with the return of summer. It was important at the start of the Celtic year (during the Festival of Samhain) for purification purposes, as its twigs were used to drive away spirits of the previous year.
- The birch was also a symbol of fertility and hence used in the festivities of Beltane, an ancient Celtic holiday celebrating the start of summer around the first of May. Beltane bonfires were lit out of birch and oak.
- In the British Isles, birch branches in or on a house were thought to bring good luck and protection from the 'Evil Eye', as did birch twigs in a buttonhole or on a hat. One Hertfordshire custom was to construct a maypole out of a tall birch tree, adorn it with red and white streamers, and place it against the stable door to protect horses against diseases and calamities.
- The Himalayan birch (*Betula utilis*) is reported to be effective for charms and spells in India. The spells were written on its bark with red ink by the priest and given to the devotee in the form of an amulet worn to cure diseases.
- The oldest known Buddhist manuscripts were written on birch bark about 1,800 years ago in Afghanistan.
- White birch bark was discovered in the tomb of Tutankhamen in Egypt, having been exported from Europe.
- The birch is one of the six national nature symbols of Finland. Two species are found there: the downy birch (*Betula pubescens*) and the European white birch (*Betula pendula = Betula verrucosa*). The birch was chosen by popular vote because of its numerous uses: the sweet sap in summer is a healthy drink; its bark is used for binding to make baskets, dishes and other containers, roofing and shoes; its timber is used for buildings, furniture, tools; its leafy twigs are made into sauna switches and, after drying, winter fodder for cattle.

Camellia japonica camellia

Scientific Species Name:
Camellia japonica
(Latin *Camellia*, after Georg Joseph Kamel or Camellus (1661–1706), Jesuit missionary and botanist; Latin *japonica*, from Japan)

Common Species Name:
camellia, common camellia, Japanese camellia (English); *camellia, camélia du Japon, rose du Japon* (French); *Japanische Kamelie, Kamelie* (German); *chá huā* [茶花], *shān chá huā* [山茶花] (Mandarin); *camelia* (Spanish)

Scientific Family Name:
Theaceae

Common Family Name:
tea family

Natural Distribution:
Korea, Japan and Taiwan

USDA 6–9 (AUS 1–3)

How to Grow: Requires more exposure in colder climates; likes warm wet summers and moderately cold dry winters; mulch soil; likes rich, moist, acidic clay or sandy soil, or slightly alkaline loam; prune after its flowering in late winter to spring; cover buds for frost protection. Propagate by semi-woody stem cuttings in late summer to winter, or air-layering in spring (blooming from 6 months), or seed but may not breed true to parent (5–20 years to bloom).

Drawbacks: Grow west of shade structure to avoid direct morning sun which withers petals; grow apart from other plants to protect its shallow roots; needs a temperature range of 5–25°C (41–77°F) to flower, so not suitable for the warm tropics; may be afflicted by viruses (no cure), fungi (apply suitable fungicides), aphids, mites, scale insects, weevils (apply suitable insecticides to eradicate).

Plant Description:

- A slow-growing, long-lived, temperate shrub or small tree 1.5–11 m (4.9–36.1 ft) tall.
- Its alternate, stalked leaves have blades which are leathery, 5–12 cm (2–4.7 in) long, 2.5–7 cm (1–2.8 in) broad, oblong-elliptic, elliptic to broadly ovate, glossy dark green above and paler below, and the margins are finely toothed.
- Its solitary, sub-terminal, rose-like flowers can grow up to 15 cm (5.9 in) across and have orange, pink, red, yellow, lavender, white or variegated petals. There are more than 3,000 cultivated varieties with varying flower forms from single (5–9 petals), semi-double to various types of double petal layers, when the numerous stamens have become petaloid. There are 9–13 sepals. The style is split into three.
- The round fruit is 3–4 cm (1.2–1.6 in) across, green when ripe with a hard covering over the compartments, each containing up to 3 or more seeds.

Camellia japonica flower and bud

 Photos: (above and opposite) Hugh Tan Tiang Wah

Camellia japonica cultivar flowers and buds

Young leaves of *Camellia japonica* cultivar resemble bronze mirrors held sacred by the Japanese

Folklore, Beliefs, Uses and Other Interesting Facts

- The shine of the camellia's glossy deep-green upper leaf blade resembles a bronze mirror, a sacred object for the Japanese, making the camellia a revered flower.
- Cultivars of camellia with large pink or red flowers are auspicious plants for the Chinese New Year. The Chinese associate red hues with auspiciousness as red is associated with blood and, hence, life. In areas with open fields, the start of a lunar year is marked by fields of revitalised plants and blooming flowers. In an urban setting, potted plants with large red flowers are commonly displayed in the home. They signify the coming of spring and bring good fortune for the coming year.
- Based on feng shui beliefs, red or pink-flowered plants are the best to grow. They are most suitable for the house front, south, southwest and northeast; bad for the west and northwest; and neutral for the rest. White-flowered plants are good for the west, northwest and north; neutral for the south, southwest and northeast; and bad for the front of the house, east and southeast.
- Based on the language of flowers, the camellia signifies pity or that beauty is one's only attraction.
- The camellia species is also a source of *tsubaki* oil, extracted from the seed. *Tsubaki* oil is used by Japanese women as a hair oil and by the Chinese as a cooking, lubricating and stamp-pad oil as well as soap.
- Considered one of the 'Queen of Plants', the camellia is the state flower of Alabama (USA); the city flower of Chongqing (China), where it has been cultivated for more than 2,000 years; and the city flower of Matsuyama City (Japan), where the camellia is thought to bring hope and happiness.
- The tea plant, *Camellia sinensis*, is a close relative.

Cananga odorata ylang-ylang

Scientific Species Name:
Cananga odorata
(= *Canangium odoratum*)
(Malay *kenanga*, the name for this species; Latin *odorata*, fragrant)

Common Species Name:
perfume tree, ylang-ylang (English); *canang odorant*, *ilang-ilang*, *ylang-ylang* (French); *Ilang-Ilang* (German); *cenanga*, *kenanga*, *kenanga hutan*, *pokok kenanga*, *semenderasa* (Malay); *yī lán* [依兰] (Mandarin); *cadmia*, *cananga*, *ilang-ilang* (Spanish)

Scientific Family Name:
Annonaceae

Common Family Name:
custard apple family

Natural Distribution:
Myanmar though the Malay Archipelago to New Guinea and Northern Queensland (Australia) and Fiji; cultivated and wild in its natural distribution, and cultivated elsewhere in the tropics.

USDA ≥11 (AUS ≥5)

How to Grow: Tropical tree; moist, well-drained soil; propagated by bud grafting, stem cuttings or seed.

Drawbacks: Not for the superstitious to cultivate; untidy looking tree when older; roots of trees may damage driveways, drains, etc.; prone to scale insects.

Plant Description:

- A tree which can grow to 30 m (98.4 ft) tall or more. Branches are horizontal to descending, the ultimate branches dangling. The dwarf form is cultivated as a pot plant.
- Alternate, stalked leaves, up to 20 cm (7.9 in) long, have oblong-ovate to oblong-elliptic leaf blades.
- The large and very fragrant flowers are arranged in axillary clusters, with the 2 whorls of 6 elongate, drooping tepals, which turn from pale green to pale yellow on maturity.
- From each flower are many fruits which turn dark purple-black on maturity.

 Photos: (opposite) Hugh Tan Tiang Wah

Mature and freshly opened flowers of *Cananga odorata*

Potted dwarf plant from a bud graft

Folklore, Beliefs, Uses and Other Interesting Facts

- This is a bad feng shui plant because of its drooping appearance which signifies sorrow, and so it is avoided.
- The plant is grown in Malay cemeteries and villages. Flowers are used for adornment or potpourri by the Malays in Southeast Asia.
- In Malay folklore of Southeast Asia, trees which emit strongly scented flowers, especially at night, are associated with ghosts, evil spirits or the *pontianak* (a female vampire who terrorises the living; originally a woman who died at childbirth and became undead).
- Ylang-ylang oil, extracted from the flowers, is used for hair-dressing and in perfumes, such as Chanel No. 5 and Revlon's Charlie.

Canavalia gladiata sword bean

Scientific Species Name:
Canavalia gladiata
(= *Canavalia ensiformis* var. *gladiata*)
(*Canavalia*, the Latinised version of the *Malabar kanavali*, vernacular name of *Canavalia* species; Latin *gladiata*, sword-like, referring to the fruit shape; Latin *ensiformis*, sword-like)

Common Species Name:
sword bean, sword jackbean (English); *dolic en sabre, haricot sabre, pois sabre, pois sabre de la Jamaïque, pois sabre rouge, pois de l'Inde, dolique sabre* (French); *Schwertbohne* (German); *kacang parang, kacang polong* (Malay); *dāo dòu* [刀豆] (Mandarin); *haba de burro, poroto sable, carabanz* (Spanish)

Scientific Family Name:
Fabaceae or **Leguminosae**

Common Family Name:
bean family

Natural Distribution:
A cultigen possibly derived from *Canavalia cathartica* (= *Canavalia virosa*), distributed in the tropics of Africa, Asia, Australia and the Pacific.

USDA >11 (AUS>5)

How to Grow: As a 'lucky' plant, only the seed is germinated to reveal the auspicious messages, and the developing plant ultimately discarded; a fast-growing climber of the humid, tropical lowlands; grows best at 20–30°C (60–86°F), 900–1,500 mm (35.4–59.0 in) per year rainfall on most soils, with pH 4.5–7; some tolerance to saline soils, waterlogging, drought, some shade; propagate by seed.

Drawbacks: Relatively hardy species; fungal diseases include root rot; insect pests include army worm, stem-boring beetle grub; should be dug up and burnt at the end of the growing season, and not kept for more than two years to avoid disease and pest problems.

Plant Description:

- Perennial climbing shrub to 10 m (32.8 ft) long but often cultivated as an annual.
- The alternate, stalked leaves each have 3 leaflets. The stalked leaflets have blades with hairs that have low density (few per square cm) and are short in length. The leaflets are egg-shaped, 7.5–20 cm (3–7.9 in) long and 5–14 cm (2–5.5 in) wide.
- Flowers resemble those of the pea, with white petals borne on an axillary, elongated shoot.
- The fruit is a linear-oblong, slightly laterally compressed pod (legume) 15–40 cm (6–16 in) long and 2.5–5 cm (1–2 in) wide, spirally-dehiscing when ripe, to release 8 to 16 seeds.
- Seeds are oblong-ellipsoid, 2–3.5 cm (0.8–1.4 in) long and 1.5–2 cm (0.6–0.8 in) wide, deep red to red brown. The seed scar (hilum) is 1.5–2 cm (0.6–0.8 in) long.

 Photo: (opposite) Abdul Aziz Agil

Canavalia gladiata beans inscribed with various auspicious Chinese idioms

Folklore, Beliefs, Uses and Other Interesting Facts

- Marketed worldwide as the 'lucky bean' or 'magic bean', the sword bean is believed to bring good fortune to its owners via inscriptions of auspicious messages made on the seed coat.
- In Asia, the sword bean is enjoying recent popularity and was extremely popular during the festive Chinese New Year season in 2006. Beliefs often arise due to commercialism and are thus cyclical in nature.
- The dormant seed is often sold in sealed containers together with dry soil. The containers range from plain tin cans to gold-painted egg shells. When watered, the seed germinates readily into a seedling. As the roots grow, the two cotyledons (seed leaves) are pushed up onto the surface exposing inscriptions denoting good wishes, lucky proverbs and even 'lucky' lottery numbers. Of course, the messages could not have appeared out of nowhere and were previously inscribed on the dry dormant seed by laser.
- This species is utilised as a cover crop for forage and green manure, and as a soil improver through increasing the nitrogen in the soil by the nitrogen fixation of the bacteria (*Rhizobium* species) in its root nodules. The sword bean can be eaten as a vegetable; its young, unripe fruits and its green seeds can be boiled; and its flowers and young leaves can be steamed. But the dry and fully matured seeds may be poisonous and should be avoided. Traditional Chinese medicine uses pink sword bean seeds.

Cardamine pratensis cuckoo flower • lady's smock

Scientific Species Name:
Cardamine pratensis
(*Cardamine*, from the Greek *kardamis*, the name of a plant in the mustard family [Brassicaceae]; Latin *pratensis*, of the meadow, referring to the habitat in which this species is found)

Common Species Name:
American cuckoo flower, cuckoo bittercress, cuckoo flower, lady's smock, may flower, meadow cress, spinks (English); *cresson des prés* (French); *Wiesenschaumkraut* (German); *cǎo diàn suì mǐ jì* [草甸碎米荠] (Mandarin)

Scientific Family Name:
Brassicaceae or **Cruciferae**

Common Family Name:
mustard family

Natural Distribution:
Europe, Kazakhstan, Russia, Mongolia, China, Korea, Japan, North America.

USDA 3–9 (AUS 1–3)

How to Grow: Damp soil of pH 5.1–7.8, as this is a plant of moist meadows and river or stream sides; propagate by seed or by dividing the root ball.

Drawbacks: May become a naturalised weed in its country of introduction.

Plant Description:

- Perennial, rhizome-bearing herb 8–80 cm (3.1–31.5 in) tall.
- The stem is unbranched, erect and bears 2 to many leaves connected to the leaf-bearing rhizome.
- The alternate, stalked, pinnate leaves have stalked, glossy, dark-green, toothed leaflets. The basal leaves (attached to the rhizome) are rosette-forming and larger than those of the erect stem. Buds may develop on the basal leaves to produce new plants.
- The flowers, with white, lilac or purple petals, are borne in an elongated flowering shoot at the tip of the erect stem.
- The narrow, elongated fruits contain many seeds.

Photos: (opposite) Henk van der Eijk

A field of *Cardamine pratensis* in a meadow at Duinlust, the Netherlands

Flowers

Folklore, Beliefs, Uses and Other Interesting Facts

- It is named the cuckoo flower in the British Isles as it flowers at the end of April when calls from migrant cuckoos from Africa begin to be heard.
- Its other common name, lady's smock, is derived from its resemblance when it is in full bloom to lady's smocks hung out to dry.
- Traditionally believed to be a fairies' plant, it is considered unlucky to remove this wild plant from its habitat and to bring it indoors. It is also believed to help generate lightning, hence it should never be brought indoors.
- In many parts of Europe, such as England and Germany, May Day (on the 1st of May) rites and celebrations mark the end of the northern hemisphere winter. Flower garlands are made to celebrate May Day, but the cuckoo flower is excluded as it is believed to be unlucky.
- However, the juice of the cuckoo plant is used to wash the eyes of babies born on May Day to prevent them from possessing the 'Evil Eye'.
- Owing to its high vitamin C content, the cuckoo flower was used in the 18th century as a remedy for scurvy.
- It is eaten in salads.

Castanospermum australe lucky bean

Scientific Species Name:
Castanospermum australe
(After *Castanea*, the generic name for the chestnut [*Castanea sativa*]; Greek *sperma*, seed, with reference to the seed which resembles that of a chestnut; Latin *australe*, Southern, referring to its natural distribution in the southern hemisphere)

Common Species Name:
Australian chestnut, black bean, lucky bean, lucky bean plant, Moreton Bay chestnut (English); *xìng yùn dòu* [幸运豆/幸運豆] (Mandarin)

Scientific Family Name:
Fabaceae or **Leguminosae**

Common Family Name:
bean family

Natural Distribution:
Coastal forests and beaches of Queensland (Australia), Western New Britain, New Caledonia and Vanuatu

USDA >10 (AUS >4)

How to Grow: When fully grown, a tropical or subtropical tree tolerant of light frost; as an auspicious plant, only the young seedling showing the two large cotyledons is grown; sow seeds by quickly germinating the large seeds at 18–25°C (64.4–77°F) after harvesting (since they die in storage) in a mix of 3:1 parts river sand and perlite; plant the seedling in loosened, moist soil of pH approx. 5.9, mulch around the base of the plant and water regularly but beware that over-watering may cause fungal root rot; many seedlings may be grown together in a pot as houseplants.

Drawbacks: Since it is grown only as a seedling, it needs to be replanted continually.

Plant Description:

- The seedling consists of 2 large, rounded to oval, dark green cotyledons 4–6 cm (1.6–2.4 in) across; dark green, white spotted young stem; and a few, young, alternate, shiny, dark green pinnate leaves with slightly curved leaflets.

Castanospermum australe seedling sold as an auspicious ornamental plant during Chinese New Year

Photos: (above) Hugh Tan Tiang Wah; (opposite) Boo Chih Min

Castanospermum australe trees along a street in Singapore

Inflorescences, unripe fruit and leaves

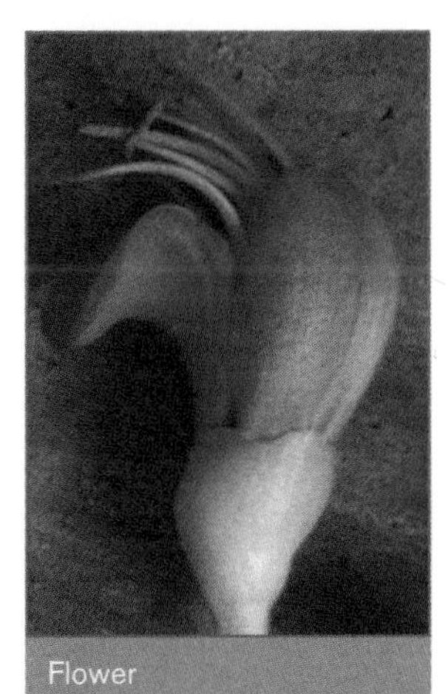
Flower

Folklore, Beliefs, Uses and Other Interesting Facts

- From popular culture in Asia, Europe and the USA, the seedling or plant is believed to bring good fortune to the owner.
- During the Chinese New Year, the Chinese purchase this auspicious plant for decoration. Each seed is sold already germinated in a small pot. The seeds are poisonous to mammals when raw but edible when roasted.
- Castanospermine, extracted from the seed, is a possible therapy to prevent HIV infection as this compound inhibits leucocyte movement to sites of inflammation, a primary means by which infection by the HIV occurs.
- It is grown as an ornamental street tree.
- The timber is attractive, resembling that of the unrelated walnut (*Juglans regia*).

Ceiba pentandra silk cotton tree • kapok tree

Scientific Species Name:
Ceiba pentandra
(Latinised South American name of this species; Latin *pentandra*, with five bundles of stamens)

Common Species Name:
copal, cotton tree, cotton wood, kapok, kapok tree, silk cotton tree, true kapok tree, white silk cotton tree (English); *arbre coton, arbre à kapok, arbre kapok, bois coton, capoc, capoquier, cotonnier de l'Inde, faux cotonnier, fromager, kapokier, kapokier de Java* (French); *Baumwollbaum, Fuma, Kapokbaum, Wollbaum* (German); *katan, safed savara, safed semul, safed simal, safed simul* (Hindi); *kabuk abu, kabu kabu, kakantrie kapok, kapok kapok, kekabu,mengkapas, pokok kapas, randu* (Malay); *jí bèi* [吉贝/吉貝], *jí bèi mián* [吉贝棉/吉貝棉], *jí bèi mù mián* [吉贝木棉/吉貝木棉], *zhuā wā mù mián* [爪哇木棉](Mandarin); *arbol capoc, arbol de seda, arbol de la seda, capoquero, ceiba juca, mosmote, peem, yuca* (Spanish)

Scientific Family Name:
Malvaceae (formerly in the *Bombacaceae*)

Common Family Name:
cotton family
(formerly in the silk cotton tree family)

Natural Distribution:
Tropical America and Tropical West Africa

USDA ≥11 (AUS ≥5)

How to Grow: Tropical tree; savanna form more drought tolerant than the rainforest form, cultivated form is a hybrid of the two; most soils; propagated by seed.

Drawbacks: Not for the superstitious; grows to a huge tree with long, thick roots which can damage driveways and drains; prickles on trunk; silk from fruiting trees can cause a mess.

Plant Description:

- A large tree which can grow up to 70 m (229.7 ft) tall. Pagoda-shaped crown in young trees with branches extending horizontally from the trunk, which can grow wide and buttressed. Sharp prickles develop on the trunk and branches.
- The stalked, alternate leaves are palmately compound each bearing 5–8 leaflets, whose blades are oblong-lance-shaped to oblong-elliptic. The blades of the leaflets grow to 20 cm (7.9 in) long.
- Flowers are large with petals that are white-based, highlighted with yellow or pink.
- Fruits, arranged in bunches, are ellipsoid capsules that grow up to 15 cm (5.9 in) long. The fruits split along 5 sutures to release the seeds, and silky hairs that grow from the inner fruit wall surface.

Photos: (opposite top and bottom) Hugh Tan Tiang Wah; (opposite middle) Angie Ng-Chua

Ceiba pentandra leafy branches and splitting fruit

Splitting ripe fruit showing the kapok fibre

Old tree along a river in Chiang Mai, Thailand

Folklore, Beliefs, Uses and Other Interesting Facts

- The silk cotton tree is revered as a spirit habitat in South America.
- In Southeast Asia, the Malays believe the tree is the abode of the *pontianak* (a female vampire who terrorises the living; originally a woman who died at childbirth and became undead).
- The silk cotton tree is grown for the silky hairs from the inside fruit wall—the kapok fibre of commerce.
- The seeds, by-products of the fibre production, can be crushed for kapok oil, and used similarly to cotton (*Gossypium* species and hybrids) oil, which it resembles. The seed cake becomes cattle food.
- There are numerous medicinal uses for its leaves, root, gum and bark.

Cercis siliquastrum Judas tree • love tree

Scientific Species Name:
Cercis siliquastrum
(Greek *kerkis*, referring to either a European species [probably a poplar {*Populus* species}], or this species, or weaver's shuttle, referring to the fruit shape; Latin *siliquastrum*, cylinder-podded)

Common Species Name:
Judas tree, love tree (English); *l'arbre de Judée* (French); *Gewöhnliche Judasbaum* (German); *nán ōu zǐ jīng* [南欧紫荆/南歐紫荊] (Mandarin); *árbol del amor*, *árbol de Judea*, *árbol de de Judas* (Spanish)

Scientific Family Name:
Fabaceae or **Leguminosae**

Common Family Name:
bean family

Natural Distribution:
Western Mediterranean to East Bulgaria, Lebanon and Turkey

USDA 5–9 (AUS 1–3)

How to Grow: Mediterranean tree that can fix its own nitrogen through its association with a bacterium growing in its roots (*Rhizobium*); deep, fertile, moist, well-drained soil of pH 5.6–8.5; drought tolerant; propagation by rooting cuttings of semi-woody stems or bud grafting of desired cultivars, or by seed.

Drawbacks: Its unfortunate association with Judas Iscariot; fungal diseases include canker, coral spot and *Verticillium* wilt; insect pests include leafhoppers and scale insects.

Plant Description:
- A deciduous shrub or tree growing to 13 m (42.7 ft) tall, with a round crown.
- The alternate, simple leaves, have broadly heart-shaped (and probably the reason for its other common name, the love tree), grey-green leaf blades up to 13 cm (5.1 in) across, frequently wider than long. Young leaves are bronze and are bright yellow when shed.
- The fragrant flowers, 1.2–1.8 cm (0.5–0.8 in) long, are pea-like, white, pale rose, purple-pink to magenta, and arranged in clusters of 3–6 on old branches or on the main trunk. Cultivated varieties vary in their colour.
- Fruits are long, compressed, sickle-shaped pods, up to 12 cm (4.7 in) long and 2 cm (0.8 in) broad, becoming dry and woody when ripe.

Photos: (opposite top) Leslie Vella; (oppposite bottom) Gabriela Bruno

Flowers of the Judas tree, University of Malta campus

Trees along a road in the village of Palmela, Portugal

Folklore, Beliefs, Uses and Other Interesting Facts

- It is generally believed that because of the confusion between *Arbor Judae* (the tree of Judas) and *Arbor Judaea* (*Cercis siliquastrum*), the latter has traditionally become the tree on which Jesus Christ's disciple, Judas Iscariot, hung himself after he betrayed his master.
- *Arbor Judaea* refers to the tree of Judaea, the southern part of ancient Palestine that was the site of the Kingdom of Judah. The tree of Judaea was frequently cultivated in the vicinity of Jerusalem, and in French is called *l'arbre de Judée*.
- A Mediterranean legend has it that the tree on which Judas hung himself is thought to be the fig (*Ficus carica*), whilst a British legend has the Judas tree as the elder (*Sambucus nigra*), and the edible fungus that grows on its trunk is Jew's ear fungus (*Auricularia auricula-judae*). Legend has it that this incident caused the previously white flowers of *Cercis siliquastrum* to become pink-red as it became stained with the shame of Judas's betrayal.
- Based on feng shui beliefs, the Judas tree (pink to magenta-flowered cultivars) would be considered auspicious because its reddish colours are symbolic of good luck and prosperity. It is, hence, very suitable for the front of the house, south, southwest or northeast of the garden; bad for the west and northwest; and neutral for the rest. The bushiness of the crown is also symbolic of prosperity.
- This species is a popular ornamental tree grown, in areas suited to it, for its beautiful flowers.

Christmas Tree

The Christmas tree is a popular tradition for the celebration of Christmas that had its origins in Europe in the 17th century. The tree is usually an evergreen conifer because of necessity, the broad-leaved deciduous trees being leafless in winter in the northern hemisphere. Over time, the species utilised changed with fashion and availability, and now also include non-conifers such as the flowering plants, pohutukawa and Western Australian Christmas tree, in the myrtle and mistletoe families respectively. Artificial trees have also become very popular, much to the horror of the traditionalists.

Scientific Species Name	Common Names	Region Used
Abies alba	European silver fir (English); *abeto* (Spanish)	Europe
Abies balsamea	balsam fir, blister fir, balm-of-Gilead fir, Canada balsam fir, eastern fir (English)	USA
Abies fraseri	Fraser fir, southern balsam fir (English)	USA
Abies nordmanniana	Caucasian fir, Nordmann fir (English); *gāo jiā suǒ lěng shān* [高加索冷杉] (Mandarin)	Europe
Abies procera	noble fir, red fir (English); *zhuàng lì lěng shān* [壮丽冷杉/壯麗冷杉](Mandarin)	USA
Metrosideros excelsa	pohutukawa, New Zealand Christmas tree (English)	New Zealand
Nuytsia floribunda	Western Australian Christmas tree (English)	Australia
Picea abies	Norway spruce (English); *epicéa commun* (French); *épinette de Norvège* (Canadian French), *gemeine fichte* (German); *ōu zhōu yún shān* [欧洲云杉/歐洲雲杉] (Mandarin); *jel europeiskaya* (Russian)	Great Britain
Picea pungens	Colorado blue spruce, blur spruce, white spruce, silver spruce, Parry spruce (English); *épinette bleue* (Canadian French); *pino real* (Spanish)	USA
Pinus sylvestris	Scotch pine, Scots pine, Baltic redwood, common pine, red pine (English); *pi blanc* [Catalan], *pi royal* [Catalan], *pinasse, pin sylvestre, pin commun, pin d'Ecosse, pin de Norvège, pin de Russie, pin de Haguenau, pin rouge du nord, pin de Riga, pin de Genève, pin d'Auvergne* (French); *Föhre, Forche, Forle, Gemeine Föhre, Gemeine Kiefer, Gewöhnliche Wald-Kiefer, Nordische Kiefer, Wald-Föhre, Waldkiefer* (German); *ōu zhōu chì sōng* [欧洲赤松] (Mandarin); *pino albar, pino blancal, pino común, pino de valsaín, pino norte, pino real, pino rojal, pino royano, pino royo, pino serrano, pino silvestre* (Spanish)	USA
Pseudotsuga menziesii	Douglas fir, Douglas yellow spruce, Douglas red spruce, Oregon pine, Douglastree (English); *de Douglas* (French); *huā qí sōng* [花旗松] (Mandarin); *piño Oregon* (Spanish)	USA
Artificial Christmas trees	Artificial Christmas trees	All countries

 Photo: (opposite) Andrew Miller

Brilliantly lit Christmas tree at Rockefeller Centre, New York City, USA

Pseudotsuga menziesii (Douglas fir) cones and leafy twigs

Abies nordmanniana (Nordmann fir) cones

Abies procera (noble fir) needle-leaved twig

How to maintain the cut Christmas tree: The tree can be purchased freshly cut from Christmas tree farms or from market lots. Prepare a tree stand that can hold water adequate enough for the tree's size. For every 2.54 cm (1 in) of the tree's diameter, the water basin of the stand should supply at least 1 litre (1.06 quarts) of water. The stand must also fit the tree perfectly. To prevent the accidental removal of the functional xylem vessels, the diameter of the trunk should not be shaved down to fit the stand. Doing this will severely compromise water transport and promote a quick drying out of the tree. When the stand is ready and filled with water, make a fresh horizontal cut 0.64–1.27 cm (¼ to ½ in) from the bottom of the trunk and put the tree on the stand. There is no need to introduce water additives, as fresh water alone is effective in maintaining the vitality of the tree. Place the tree in a cool spot away from direct heat sources to help extend its lifespan. Check the water level in the stand regularly and ensure that the cut end of the trunk is always submerged in water.

Drawbacks: A major drawback of Christmas trees is the likelihood of fires. In the USA, between 2002 and 2005, there was an estimated average of 210 fires per year from ignited natural or artificial Christmas trees, causing an average of 24 deaths, 27 injuries and US$13.3 million in property damage per year.

If you intend to get an artificial tree, ensure that is is made of fire-retardant material. Never use lit candles to decorate your Christmas tree. If your tree is decorated with electric light bulbs, take the following precautions to prevent fires from occurring: [1] use electric lights listed by a reputable testing laboratory; [2] follow the manufacturer's instructions on their use; [3] do not use the lights if the wires are worn, frayed or broken, or there are loose bulb connections; [4] do not overload the power source; [5] avoid placing the tree such that the power cord runs long distances. [6] always unplug the lights before leaving home or going to bed; [7] keep the tree as moist as possible by replenishing the water daily; [8] use a sturdy tree stand to prevent it from toppling over; [9] keep an eye on children around Christmas trees; [10] ensure that the trees are at least 1 m (3.3 feet) from any heat source, such as radiators or fireplaces; [11] safely dispose of the tree when it starts to shed needles as a dried tree is highly flammable, and do not place it in a garage or propped against the house.

Other drawbacks of natural Christmas trees include: possible cause of allergies; the opportunity cost of the land and resources utilised for tree farms (even though farms provide habitats for wildlife and soil erosion protection and other environmental services); fertiliser or pesticide impacts unless organically grown;

Photos: (this page, top to bottom) Steven Patch; Bill Strong; Hugh Tan Tiang Wah; (opposite page, top to bottom) Peter Birch; Werner Drizhal

Pinus sylvestris (Scots pine) cones on leafy branch

Flowers of *Metrosideros excelsa* (pohutukawa)

the release of carbon dioxide when disposed by incineration; landfills do not decompose and take up space. On the other hand, the advantages include: disposal of the used trees by composting or mulching; as erosion barriers; or the creation of habitats for aquatic organisms when sunk into water bodies. One more way to be more environmentally friendly is to use energy saving lights for the tree, such as the LEDs (light-emitting diodes) used on the 2007 Rockefeller Center Christmas tree.

Plant Description:

- In the northern hemisphere, Christmas trees are usually conifers of the pine family (*Pinaceae*) trimmed into a conical form.
- Leaves are simple, linear to needle-shaped.
- Species of the pine family are non-flowering and instead have separate male and female cones borne on each plant.
- In the southern hemisphere, non-conifers are used instead. In Australia, the native gregariously flowering tree, the Western Australian Christmas tree (*Nuytsia floribunda*), is used. In New Zealand, the spreading coastal forest tree, the New Zealand Christmas tree (*Metrosideros excelsa*), is regarded as the country's Christmas tree.
- Artificial trees will resemble natural trees in the conical form of the crown, branches and leaves.

Abies balsamea cones

Photos: (above) Winfield Sterling; (opposite top) Ruth Ellison; (opposite bottom) Chrissie Jamieson

Nuytsia floribunda (Western Australian Christmas tree), Perth, Western Australia

Nuytsia floribunda buds and flowers, Western Australia

Folklore, Beliefs, Uses and Other Interesting Facts

- The tradition of having a tree in the house to celebrate Christmas originated in Germany. The first known Christmas tree record is dated 1605, when decorated fir trees were placed in the parlours of Strasburg, Germany.
- From Germany, the Christmas tree was introduced to other parts of Europe and North America. It was not known in the British Isles until 1840 when Prince Albert, the husband and consort of Queen Victoria of the United Kingdom of Great Britain and Ireland, brought it to the court of St. James. This practice permeated through the aristocracy and the wealthy merchant class to the whole of London.
- The early Christmas tree in insular Sweden and Baltic Russia was a fir decorated with nuts and apples and carried five candles on each branch.
- While the spread of the Christmas tree is well documented, the origin is uncertain. Tales relating the tree to historical figures are well known. One of the more popular ones starts with Martin Luther wandering across the countryside and being impressed by the bright starry sky. Upon reaching his home, he immediately cut a fir tree, placed it indoors and dressed it with candles to reflect the heavens in all of its nocturnal glory, with stars shining as brightly as when Jesus descended to earth. According to A. Tille, this legend is probably derived from a picture by Schwerdtgeburth titled 'Luther Taking Leave of His Family', which shows Luther's family around a Christmas tree.
- Another charming legend, also recorded by Tille, is set in Lindenau, Germany. During the Thirty Years' War between Germany and Sweden, an injured Swedish officer was nursed with great kindness by the Protestant villagers. His wound was healed by Christmas time, but before leaving for Sweden, he organised a Christmas festival, with the help of the clergyman, to thank the people. The festival was similar to that celebrated in his homeland and, for the very first time, a fir tree decorated with lights was placed in the church.
- The commercial production of noble and Fraser firs in North America and Nordmann firs in Europe has been increasing in recent years. The noble and Fraser firs are considered to have the best post-harvest (best keeping) quality among the other conifer species commercially grown for Christmas. When harvested and displayed at home in water at 20°C (68°F), these species can last for at least 6 weeks with very minimal needle loss.

Cibotium barometz Scythian lamb • woolly fern

Scientific Species Name:
Cibotium barometz
(Greek *kibotos*, a small box, referring to the sorus [sporangial cluster] of this species; Tartar *barometz*, lamb, referring to the woolly tip of the trunk or terminal bud)

Common Species Name:
golden rooster, Scythian lamb, Tartarian lamb, woolly fern (English); *bulu pusi, bulu empusi, penawar jambi* (Malay); *jīn máo gǒu* [金毛狗] (Mandarin)

Scientific Family Name:
Dicksoniaceae

Common Family Name:
Australian tree fern family

Natural Distribution:
India to Southern China, to Malaysia and Indonesia

How to Grow: As an auspicious plant or botanical curiosity, usually only grown as the stem tip for its development of one coiled leaf (golden chicken), or not at all; as an exhibition piece, with the structure inverted, with four leaf stalks cut short to resemble legs of a quadruped (Scythian or Tartarian lamb); keep indoors and water when the pot soil becomes dry; subtropical to tropical fern.

Drawbacks: Will grow into an 'ugly duckling', a large, rather untidy-looking fern.

Plant Description:
- A large fern with a short, prostate stem that is covered with a thick layer of shiny, golden brown hairs. The tip of the stem is covered with long, golden brown hairs and, when excised, becomes the golden rooster or Scythian lamb of commerce and legend.
- The spirally arranged, stalked, bipinnate leaves grow up to 2 m (6.6 ft) long. Its leaflets have glossy and glabrous upper sides and fine whitish bloom-covered undersides.
- Leaflets are broadly ovate-lanceolate. Secondary leaflets are linear and segments linear oblong with a pointed tip.

Cibotium barometz shoot tip in vase sold commercially

Photos: (above) Hugh Tan Tiang Wah; (opposite top) Giam Xingli; (opposite bottom) Wee Yeow Chin

Cibotium barometz plant

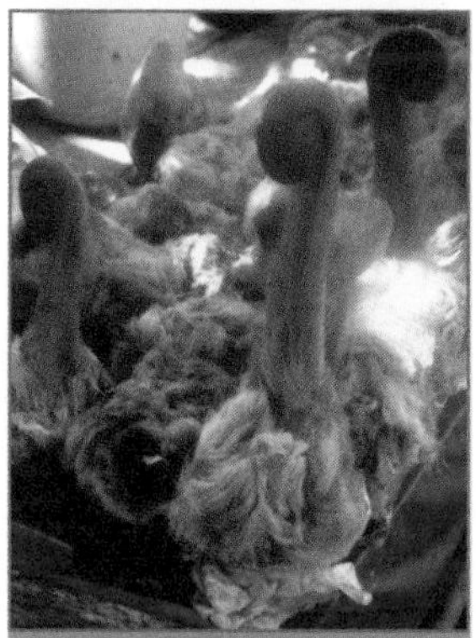

Shoot tips on sale, each sprouting a leaf

Folklore, Beliefs, Uses and Other Interesting Facts

- *Cibotium barometz* is closely linked to the fable of the 'Vegetable Lamb of Tartary' or Scythian lamb, hence its common name.
- The fable originates from Central Asia and describes the Scythian lamb as a plant which grows a sheep as its fruit, or as a 'lamb' rooted to one spot, and would die when it had eaten all plants within its reach. It has been suggested that the fern could be the plant which inspired this fable because if one were to invert the hairy, stout stem tip which resembles the body of a lamb, the cut leaf stalks will resemble its legs, and hence was considered half animal and half plant. News of the Scythian lamb perhaps reached Europe because the Chinese were importing it for medicinal use, although it also grows in China.
- In Southeast Asia, the Scythian lamb has been a traditional Chinese New Year plant. The hairy prostate stem tip, shaped like the body of a headless chicken, contributes to the enigma of this odd-looking plant especially when it is sold undeveloped and leafless, so one cannot be blamed for questioning its place in a plant nursery. Over time, the new leaf emerges, and, being a fern, the young leaf becomes a 'fiddlehead', or 'head of the chicken', hence the 'golden chicken' name.
- The Chinese in Southeast Asia believe that golden eggs are a symbol for wealth and prosperity. The golden rooster plant, believed to be able to 'bear golden eggs', is a favourite during the Chinese New Year period as an auspicious symbol of wealth and prosperity. Its popularity peaked during the Chinese New Year of 2004, the lunar year of the rooster.
- Besides being grown as an ornamental plant, the hairs have also been used medicinally by the Chinese for centuries as a styptic (substance to stop bleeding) for dressing wounds. The hairs have also been used for stuffing pillows, but they have been reported to escape the pillow case and get into the lungs, posing possible danger.

×Citrofortunella microcarpa

limequat • calamondin

Scientific Species Name:
×Citrofortunella microcarpa
(= *Citrus madurensis*, *Citrus microcarpa*, *Citrus mitis*, ×*Citrofortunella mitis*)
(× = multiplication sign, signifying an intergeneric hybrid status; *Citrofortunella* from combining the generic names, *Citrus* and *Fortunella*; Greek *micros*, small; Latin *carpus*, fruit, referring to the small fruits)

Common Species Name:
calamandarin, calamondin, calamondin orange, calamonding, China orange, Chinese orange, golden lime, kalamansi lime, limequat, musk lime, Panama orange, Philippine lime, scarlet lime (English); *calamondin* (French); *Zwergapfelsine* (German); *limau cuwit* (*chuit*), *limau kesturi* (Malay); *gān* [柑], *jīn jú* [金橘], *sì jì jú* [四季橘], *yuè jú* [月橘] (Mandarin); *naranjita de San José* (Spanish)

Scientific Family Name:
Rutaceae

Common Family Name:
citrus family

Natural Distribution:
Originated in China as a natural hybrid of probably *Citrus reticulata* (Mandarin orange) and *Fortunella margarita* (oval kumquat); now cultivated especially in east, southeast and south Asia and also elsewhere

USDA ≥11 (AUS ≥5)

How to Grow: Subtropical to tropical shrub or tree; frost intolerant; sandy-loam, organic rich, slightly acid to slightly alkaline [pH 5.5–7.0], well-drained soils; medium tolerance for drought; intolerant of strong winds; propagate by seed (but these do not breed true) or stem cuttings, air-layering and grafts.

Drawbacks: Leaf mottling disease; crinkly leaf, exocortis, psorosis and xyloporosis viral diseases; attracts butterflies which feed on nectar and their caterpillars, on leaves; sooty moulds on leaves and scale insects.

Plant Description:

- An evergreen shrub or small tree to 7.5 m (24.6 ft) tall.
- The stalked, alternate leaves possess elliptic to obovate, 3–8 cm (1.2–3.1 in) by 1–4 cm (0.4–1.6 in) leaf blades which have shallowly toothed margins.
- Fragrant flowers occur singly or in small clusters up to 3 in the leaf axils. Each flower has 5 white petals that are 1–2 cm (0.4–0.8 in) long, and numerous stamens with their filaments joined into a tube.
- The fruit is round, 2–4 cm (0.8–1.6 in) wide, yellowish green to reddish orange when ripe, with numerous oil glands in the thin rind which surrounds the 6–10 segments containing the orange, juicy flesh.
- Each fruit has 0–11 egg-shaped seeds.

 Photos: (opposite) Hugh Tan Tiang Wah

Dwarf ×*Citrofortunella microcarpa* plants for sale during Chinese New Year

Flowers, leaves and fruits

Folklore, Beliefs, Uses and Other Interesting Facts

- To the Chinese, the limequat is an important auspicious Chinese New Year ornamental. Like the Mandarin orange or kumquat, its orange fruits symbolise gold coins that signify an abundance of wealth. From a feng shui standpoint, a pair of heavily fruiting plants, placed at the front door of the house, is extremely beneficial as they symbolise general prosperity, great good fortune, happiness and wealth. The Cantonese name of this plant, *kum*, is a phonetic pun for 'gold' (Mandarin *jīn* [金]). Greenhouse grown plants can be made to fruit during winter and can be brought into the house to, symbolically, bring in luck. As the limequat does not drop its fruit through winter, it is a very good plant for this purpose.
- Its main economic value is its refreshing, acidic fruit juice that is drunk fresh, bottled or diluted from the bottled concentrate. Its freshly squeezed juice is a flavour enhancer for seafood or meat, a deodorant, stain remover, skin bleach and shampoo for hair.
- The fruit is preserved in syrup or made into marmalade or chutneys.
- Medicinally, the juice is used for treating skin irritations, coughs, laxative, phlegm expulsion (with pepper) or reducing inflammation. The oil, distilled from the leaves, is used to treat flatulence.
- In horticulture, it is used as a rootstock for lemons (*Citrus limon*) and as a graft for oval kumquats (*Fortunella margarita*).

Citrus medica var. sarcodactylis

Buddha's hand • Buddha's hand citron

Scientific Species Name:
Citrus medica* var. *sarcodactylis
(Latin *citrus*, citron [*Citrus medica*]; Latin *medica*, healing or medicinal; var. = variety, a subgroup of the species; Greek *sarkos*, fleshy; Greek *daktylos*, a finger)

Common Species Name:
Buddha's hand, Buddha's hand citron, fingered citron, flesh finger citron (English); *cédrat digité*, *cédrat main de Bouddha*, *main de Bouddha*, *sarcodactyle* (French); *Buddhafinger*, *gefingerte Zitrone* (German); *limau jari*, *limau kerat lintang* (Malay); *fó shǒu* [佛手], *fó shǒu gān* [佛手柑], *wú zhǐ gān* [五指柑] (Mandarin)

Scientific Family Name:
Rutaceae

Common Family Name:
citrus family

Natural Distribution:
Origin uncertain, possibly India and introduced into China by Buddhist monks; now widely cultivated in China, India, Indo-China and Japan

USDA ≥9 (AUS ≥3)

How to Grow: Tropical shrub or tree; sensitive to frost and intense heat or drought; most delicate of the cultivated *Citrus* species; moist, well-drained, deep and fertile soils preferred; propagation by leafy stem cuttings or bud grafted onto rootstocks of other *Citrus* species.

Drawbacks: Straggly, untidy appearance of the plant; thorny stems and branches; branch knot (fungal disease); mites and scale insect attack.

Plant Description:

- A small tree of up to 3–5 m (9.8–16.4 ft) tall. Young twigs are purplish-brown. Axillary thorns are found on branches and stems. Often grown as a dwarf ornamental plant.
- Alternate, stalked leaves have leaf blades 5–20 cm (2.0–7.9 in) by 3–9 cm (1.2–3.5 in), are elliptic-ovate to ovate-lanceolate with saw toothed margins. Leaves are fragrant when crushed.
- The bisexual or male flowers grow in few-flowered clusters. They have 5 white petals which are pinkish outside, many stamens and a thick style on top of the ovary.
- Fruits are light yellow when ripe, growing to 20 cm (7.9 in) long. The bizarre looking fruit arises from the failure of the carpels of the ovary to unite and the carpels develop individually, resembling a hand with fingers clasped together. This is also the reason why many Chinese, in particular the Buddhists, refer this fruit as Buddha's hand, or *fó shǒu*. The fruit is very fragrant.

Side view of the *Citrus medica* var. *sarcodactylis* fruit

Photos: (above and opposite) Hugh Tan Tiang Wah

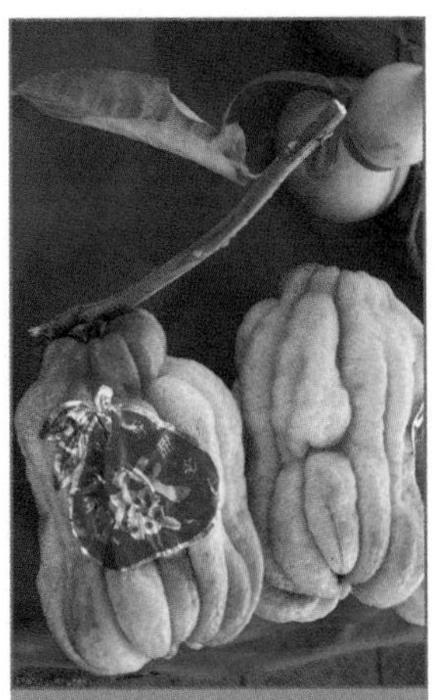

Citrus media var. *sarcodactylis* fruits for sale at Chinese New Year

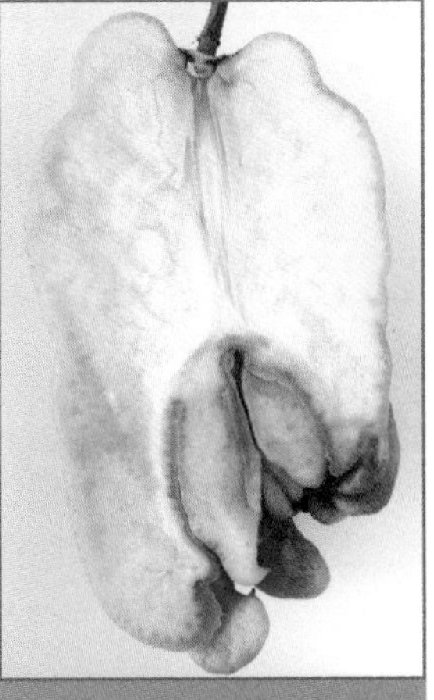

Half section of fruit

Folklore, Beliefs, Uses and Other Interesting Facts

- Buddhists regard the Buddha's hand citron to be one of the sacred fruits of Buddhism, the presiding factor being the imagery of the fruit as the hands of Buddha.
- Another factor which may have contributed to the use of the Buddha's hand citron as an altar offering at home or in temples is the fragrance of the fruit, which is also used to perfume rooms and clothes.
- During the Chinese New Year period, it is also used as an ornamental fruit to ensure the well-being of the family which displays it.
- Based on feng shui principles, this plant is not considered auspicious for the garden. The numerous thorns on the stem and branches represent 'poison arrows' emitting hostile, inauspicious energy, and so it is best placed as far away from the house as possible and near the gate, to be used as a friendly sentinel.
- The Chinese use its flowers medicinally. Fresh twigs are boiled as a decoction for a stomachache, appetite enhancer or for deworming. The leaves are used as a poultice for aching joints.

Citrus reticulata mandarin orange • tangerine

Scientific Species Name:
Citrus reticulata
(Latin *citrus*, the citron [*Citrus medica*]; Latin *reticulata*, with a net-like pattern, possibly referring to the appearance of the inner surface of the fruit wall [peel])

Common Species Name:
common mandarin, culate mandarin, mandarin, mandarin orange, suntara orange, tangerine, true mandarin (English); *mandarine*, *mandarinier* (French); Mandarine, Mandarinenbaum (German); *santara* (Hindi); *limau langkat*, *limau kupas*, *limau wangkang* (Malay); *gān* [柑], *jié* [桔], *jú* [橘], *mì gān* [蜜柑], *kuān pí gān* [宽皮柑], *kuān pí jié* [宽皮桔], *pèng gān* [碰柑], *tū gān* [凸柑] (Mandarin); *mandarina*, *mandarino* (Spanish)

Scientific Family Name:
Rutaceae

Common Family Name:
citrus family

Natural Distribution:
Southeast Asia

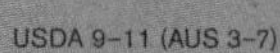

USDA 9–11 (AUS 3–7)

How to Grow: Tropical shrub or tree so not frost tolerant; tolerates soils that are clayey, loamy, sandy, slightly alkaline or acidic, well-drained; moderately drought-tolerant; low tolerance for salt aerosols; propagated by bud grafting or air-layering.

Drawbacks: Prone to pests such as nematodes, scales, whiteflies, mites, caterpillars, fruit flies, greening-virus complex and fungal diseases; often afflicted with sooty moulds.

Plant Description:

- Thorny shrub or tree with slender twigs.
- The alternate leaves, up to about 4 cm (1.6 in) long, possess elliptic to lanceolate leaf blades and winged leaf stalks, bearing thorns in their axils.
- The fragrant flowers arising singly or in small clusters in the leaf axils possess waxy white petals.
- The fruit is nearly round to round-flattened and about 8 cm (3.1 in) in diameter. Its peel is characteristically thin and loose, separating easily from the 10–14, equally loose, segments. The pulp is fragrant and very sweet, housing the few seeds.

Citrus reticulata 'Swatow' whole fruit and dismantled to show the segments

Photos: (above and opposite) Hugh Tan Tiang Wah

Citrus reticulata plants for sale in a nursery during Chinese New Year

Citrus reticulata fruits are used in lion dances to symbolise prosperity

Folklore, Beliefs, Uses and Other Interesting Facts

- The mandarin orange is an auspicious fruit for the Chinese, especially during the Chinese New Year. This is the time when the Chinese visit the homes of friends and relatives to wish each other good luck for the new year and to catch up on the last year. The Chinese custom of house-visiting includes the exchange of a pair of mandarin oranges between the host and the visitor. The golden-orange colour of mandarin oranges resembles the golden colour of coins, and receiving them during New Year bodes incoming wealth and fortune for the coming year. The Cantonese name of the mandarin orange, *kum*, is a phonetic pun for gold, hence its association with prosperity.
- Feng shui belief is that a pair of heavily fruiting mandarin orange trees is highly beneficial as they signify good fortune, great prosperity, wealth and happiness. The orange fruit symbolises abundant gold. A pair of potted plants should be placed on either side of the front door or, if planted in the ground, should be placed within sight of the front door or, if the garden is behind the house, then at the back door.
- There are numerous cultivars commercially grown for their fresh fruit, with exports going into the millions of tons annually worldwide. China, Taiwan, Thailand, Pakistan, USA and Australia all export mandarin oranges. Mandarin orange oil is also used in shampoos and cosmetics.
- In the language of flowers, the orange blossom signifies chastity, generosity, magnificence or that one's purity equals one's beauty.
- Chinese farmers have used a leaf-nesting ant (*Oecophylla smaragdina*) since AD 304 for biological control against insect pests, with bamboo poles between trees to allow movement of the ants. The leaf-nesting ants are used to keep the mandarin orange plants insect-free without the use of pesticides.

Cocos nucifera coconut • coco palm

Scientific Species Name:
Cocos nucifera
(Portuguese *coco*, mask;
Latin *nucifera*, nut-bearing)

Common Species Name:
coco palm, coconut, coconut palm, nariyal (English); *noix de coco, cocotier* (French); *Kokos, Kokosnuß, Kokosnuss, Kokospalme* (German); *naariyal* (*nariyal*), *naariyal kaa per, khopar* (Hindi); *kelambir, kelapa, kerambil, nyiur* (Malay); *yē zǐ* [椰子], *yē shù* [椰树] (Mandarin); *coco, cocotero, nuez de coco, palma de coco, palmera de coco* (Spanish)

Scientific Family Name:
Arecaceae or **Palmae**

Common Family Name:
palm family

Natural Distribution:
Now pan-tropical and widely cultivated, but natural distribution uncertain (possibly originated in the Western Pacific)

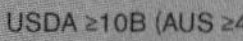

USDA ≥10B (AUS ≥4)

How to Grow: Tropical palm; not frost tolerant; well-drained soil; high drought and salt tolerance; regular fertiliser application; propagated by seed (half bury brown fruit in soil); seed becomes unviable after a fortnight in seawater.

Drawbacks: Falling leaves and fruits are dangerous; the growing trunk tip is prone to rhinoceros beetle attack, which will kill the tree; susceptible to lethal yellowing disease (virus), but dwarf cultivars such as 'Golden Malayan Dwarf' are resistant; other viral diseases, fungal diseases and nematodes.

Plant Description:

- A solitary-trunked, erect palm that grows to 20–30 m (65.6–98.4 ft) tall. The uniformly slender trunk is enlarged basally.
- The alternate, pinnate leaves, with up to 100 leaflets, are bunched at the trunk tip.
- The fragrant, stalkless flowers are arranged on a shoot, with a single, woody female flower at the base, and many smaller male ones above. These flowering shoots branch off a main shoot.
- The coconut fruit is not a nut but rather a fibrous drupe, with the seed surrounded by a hard, tough, final layer of the fruit wall. The layer just above this is a thick, fibrous, air-filled zone (for buoyancy) and above this is the 'skin' which waterproofs it. The coconut is water-dispersed.
- The seed is round and, in the early stages, consists mostly of 'coconut water' (liquid endosperm) which everyone enjoys as a refreshing drink. Over time, the seed forms a hard white layer (solid endosperm), from which 'coconut milk' can be squeezed.

 Photos: (opposite top) Giam Xingli; (opposite middle and bottom) Hugh Tan Tiang Wah

Cocos nucifera tree

Fruits in the tree crown

Inflorescence between two leaf stalks

Folklore, Beliefs, Uses and Other Interesting Facts

- Hindus regard the coconut fruit as sacred, and offer it to their gods in temples. The three marks or 'eyes' on its surface represent the three eyes of Lord Siva.
- Hindus also believe the fruit's 'eyes' represent the three eyes of humans, the third eye being invisible but able to differentiate between true and false. The different parts of humans are also compared to the fruit and seed layers. The fibrous husk represents the physical human, while the solid and the liquid endosperms represent the psychological and the spiritual aspects respectively. Hence, rituals that involve the breaking of the coconut seed before deities symbolise the destruction of the human ego and the submission of the body for purification. (In ancient India, human sacrifice was used to propitiate the deity, but this was replaced by a coconut, which superficially resembles a human head.)
- The broken pieces of the coconut endosperm are presented to Hindu devotees for consumption as *Prasadh*, a blessed holy food, to nourish them with blessings from the deity.
- In the Hindu festival of Thaipusam, the Kavadi carrier smashes the coconut seed (fruit wall removed) against the ground en route to their final destination. This signifies the breaking of his ego and his full submission to the gods so that the purification process can be completed by the end of the trip.
- To bless and protect a 30-day-old newborn from harm, Malays place the baby's hair into the endosperm of the coconut fruit.
- Based on feng shui belief, trees with long, narrow, single trunks, such as the coconut or other similar palms, represent 'poison arrows' which emit hostile energy, so should be avoided in gardens and must be blocked off if they face the main door.
- The coconut is a very useful plant. Oil from the endosperm is used for cooking and for soap; its husk is used for rope, matting and as a growing medium; and its fruit is eaten. The sugary phloem sap is also fermented to become toddy. To tap this, the top of the inflorescence is cut, leaving only the inflorescence stalk, and the phloem sap which flows out from the cut tip is collected.

Conium maculatum poison hemlock

Scientific Species Name:
Conium maculatum
(Latin *conium*, this species; Latin *maculatum*, spotted, referring to the characteristic purple-spotted stem and branches)

Common Species Name:
bad-man's oatmeal, beaver poison, bunk, cashes, Devil's oatmeal, heck-how, hemlock, poison hemlock, poison parsley, poison root, poison snakeweed, St.-Bennet's herb, spotted conium, spotted cowbane, spotted hemlock, spotted parsley, wode-whistle (English); *ciguë* (French); *Gefleckter Schierling*, *Schierling* (German); *shaukaran* (Hindi); *hemlok* (Malay); *dú shén* [毒参] (Mandarin); *cicuta* (Spanish)

Scientific Family Name:
Apiaceae or **Umbelliferae**

Common Family Name:
celery family

Natural Distribution:
Mediterranean but a naturalised weed in the North temperate regions of Asia, Europe and North America

USDA 5–10 (AUS 1–4)

How to Grow: Prefers soggy soil of pH 5.6–7.5; propagation by dividing rhizomes, stem cuttings or seed.

Drawbacks: All parts are highly poisonous; may become an invasive weed.

Plant Description:

- A biennial herb whose leaves resemble those of parsley (*Petroselinum crispum*) to 0.6–3 m (2.0–9.8 ft) tall but, unlike parsley, has an unpleasant, mouse-like odour when crushed.
- It has a smooth, hollow, purple-spotted, much-branched stem.
- It has stalked, bi- to tripinnate leaves up to 30 cm (11.8 in) long and wide.
- It produces flat-topped clusters of tiny white flowers in mid-summer.
- The fruit, which ripens in late summer, is greyish green, longitudinally ribbed, dried when ripe and 2–4 mm (0.1–0.2 in) long.

Conium maculatum fruits

 Photos: (above) Steve Hurst @ USDA-NRCS PLANTS Database; (opposite) Jaap Cost Budde

Conium maculatum flowers and young fruit clusters

Flowering plant

Folklore, Beliefs, Uses and Other Interesting Facts

- Poison hemlock contains very poisonous alkaloids (polyacetylenes, including coniine, considered the most poisonous) and was the official poison used in executions in ancient Athens. The Greek philosopher, Socrates, was executed for impiety against the Greek gods in 399 BC using a solution of poison hemlock. Poison hemlock has been used for killing rodents since the 2nd century AD.
- In the UK, poison hemlock was associated, by tradition, with the devil and his flock and the witches who are reported to utilise poison hemlock for spells to invoke evil spirits and demons. Children in the northern counties of England are told to avoid touching the plant because the devil may snatch them away if they do. The plant is also believed to demolish love, induce insanity or paralysis and cause sterility in humans and animals.
- Hemlock was sacred to Hecate, the Greek goddess of the underworld and protector of witches.
- According to German folk tradition, the hemlock was home to a toad, which lived beneath it and sucked up its poisons.
- In the language of flowers, poison hemlock, not surprisingly, symbolises bad conduct or says "You will cause my death".
- Plant extracts have been used medicinally to relieve pain and may be a cancer remedy.

Cordyline fruticosa ti plant

Scientific Species Name:
Cordyline fruticosa
(= *Convallaria fruticosa*, *Cordyline terminalis*, *Dracaena terminalis*) (Greek *kordyle*, a club, referring to the club-like, swollen, fleshy roots of *Cordyline* species; Latin *frutex*, a shrub)

Common Species Name:
false palm, good luck plant, ki, palm lily, Polynesian ti plant, ti, ti plant, tree of kings (English); *Keulenlilie* (German); *jeluang*, *sawang* (Malay); *zhū jiāo* [朱蕉] (Mandarin); *caña de indio*, *croto* (Spanish)

Scientific Family Name:
Laxmanniaceae
(formerly placed in the *Agavaceae*)

Common Family Name:
paperlily family
(formerly in the *agave* family)

Natural Distribution:
Unknown, but possibly ranging from the Himalayas, Southeast Asia and New Guinea to Northeast Australia. Spread widely by humans to Melanesia, and Polynesia to Hawaii where it was frequently cultivated.

USDA 10B (AUS ≥4)

How to Grow: Tropical to subtropical plant; slightly alkaline to acidic, clay, loam, sand or fertile, well-drained soil; weakly drought-tolerant but intolerant of temperatures below 13°C (55°F); propagation by stem cuttings (place in a bucket of tap water until roots appear).

Drawbacks: Generally disease and pest-free; may be afflicted with leaf spot disease, nematodes, mealy bugs, mites and fluoride damage; poor tolerance for salt and best not grown near the sea.

Plant Description:

- Unbranched to rarely branching tropical to subtropical, palm-like, perennial plant, growing up to 3–4 m (9.8–13.1 ft) tall. The lower end of the slender stem is conspicuously marked by leaf scars. Suckers may be produced at the stem's base. The roots become swollen and fleshy.
- The large, stalked leaves are crowded at the stem tip, ranging from green to red, or variegated. The leaf blades range from linear to oblong and lance-shaped. Colours are brighter in sunnier positions.
- Sweetly fragrant, yellowish or reddish white flowers, about 1 cm (0.4 in) across, grow in clusters at the tips of mature plant stems.
- The fruits are red berries.

Cordyline fruticosa inflorescence growing from the tip of the stem

Photos: (above and opposite top) Hugh Tan Tiang Wah; (opposite bottom) Erwyn Arizo

Leafy stems

Hula skirt made of *Cordyline fruticosa* leaves, Lahaina, Hawaii

Folklore, Beliefs, Uses and Other Interesting Facts

- In ancient Hawaii, the ti plant was regarded to have certain divine powers, and only tribal chiefs and spiritual leaders were allowed to wear a necklace of its leaves during rituals and ceremonies. The spiritual property of the ti plant has not been lost over time as many Hawaiians plant it as a hedge around their homes to ward off evil and for good luck. The plant is also recognised to be a good luck plant in the USA and India.
- Malays in Southeast Asia believe that the ti plant wards off evil and grow it at the edges of graveyards and at the corners of the house to prevent evil spirits from leaving their abode and entering their homes. Malayan tribes hang red-leafed ti plants over the head of women in confinement to protect them from evil spirits. This practice has been recorded since the early 20th century.
- Based on feng shui beliefs, it would be auspicious to plant the red-leaved cultivars, whose colour and posture resemble fire, in the south, southwest or northeast of the garden, or the left of the front of the garden. Red is considered auspicious, symbolising the crimson phoenix, one of the four celestial creatures that are the mainstays of classical feng shui. Planting this plant in the north has a neutral effect, but it would be inauspicious to plant it in the west, northwest, east or southeast.
- Ti plants are used as ornamental plants for hedges in the subtropics and tropics; for traditional dress in New Guinea; for hula skirts in Hawaii; for wrapping food for cooking, roof thatch, rain coats, sandals, plates, food wrappers, cattle fodder and soap (after crushing) in the Amazon; and its sweet, starchy roots are eaten in Oceania.

Crassula ovata jade plant • friendship tree

Scientific Species Name:
Crassula ovata
(Latin *crassus*, thick, referring to the thick, fleshy leaves found in *Crassula* species; Latin *ovata*, egg-shaped with the broader end basal, referring to the leaf blades of this species which are actually obovate, with the broader end towards the tip instead)

Common Species Name:
baby jade, cauliflower-ear's, dollar plant, friendship tree, jade plant, jade tree, Japanese rubber plant, Japanese laurel, money tree (English); *yù shù* [玉树/玉樹] (Mandarin)

Scientific Family Name:
Crassulaceae

Common Family Name:
jade plant family

Natural Distribution:
Mozambique and South Africa

USDA ≥10 (AUS ≥4)

How to Grow: No direct sunlight; desert plant; well-drained soil (3:1:1 compost-grit-coarse sand); water only when necessary; propagate by leaf or stem cuttings.

Drawbacks: Not easy to grow in the humid tropics and certainly not outdoors; mealy bug infestation may cause growth deformation; aphids may feed up flower stalks; over-watering will lead to rotting stems and leaves.

Plant Description:

- A succulent perennial that can grow to more than 1 m (3.3 ft) tall. The thick, fleshy stem branches frequently.
- The opposite, stalked leaves have thick, fleshy, elliptic to obovate, 2–4 cm (0.8–1.6 in) by 3 cm (1.2 in), shiny, jade green leaf blades whose margins may be tinged purple-red.
- The star-like white flowers usually have 5 petals clustered at the branch tips. Flowering is believed to be induced by long night periods.

Portulaca afra (miniature jade) plant

Photos: (above and opposite bottom) Boo Chih Min; (opposite top) Hugh Tan Tiang Wah

Potted plants of *Crassula ovata* at a nursery for sale during Chinese New Year

Leafy branches of *Portulaca afra*

Folklore, Beliefs, Uses and Other Interesting Facts

- The jade plant is gaining popularity as an ornamental plant in many countries, especially in North America, Europe and Asia. The shiny, green leaves resemble pieces of valuable jade and represent an abundance of the precious stone, indicating plentiful wealth at that home or blessing it with prosperity.
- According to feng shui belief, a pot of the jade plant should be placed inside or outside the front door to invite prosperity and abundance. The jade plant is a typical auspicious feng shui plant for prosperity: succulent, with thick leaves with water-filled tissues; shiny, round and plump (like coins); and dark green (like paper money or jade). Other plants with similar features include *Portulaca afra* (also called the jade plant, or miniature jade plant, purslane tree, spekboom, elephant's food) which has smaller, proportionately broader leaves and leaf blade tips that are blunter or even notched. In Bangkok, even mangrove trees (e.g., *teruntum–Lumnitzera* species), which have thick, fleshy bright green leaves, are being sold as auspicious feng shui plants.
- The jade plant has white flowers, which are usually not considered auspicious, as white flowers are used for funerals and mourning based on traditional Chinese belief. This will not pose a problem if it does not flower. If it does flower, it can be replaced with a non-flowering jade plant.
- The jade plant's main economic value is as an ornamental pot plant, and is thus grown for its attractiveness and/or auspiciousness.

Crataegus species hawthorn • thorn apple

Scientific Species Name:
***Crataegus* species**
(Greek *crataegus*, hawthorn; Greek *kratos*, strength, referring to the hardness and strength of the wood of this genus; there are more than 1,000 *Crataegus* species and many cultivated varieties)

Common Species Name:
haw, hawthorn, thorn, thorn apple, whitethorn (English); *shān zhā* [山楂] (Mandarin)

Scientific Family Name:
Rosaceae

Common Family Name:
rose family

Natural Distribution:
Northern temperate

USDA 4–9 (AUS 1–3)

How to Grow: Temperate tree or shrub; medium rate of growth; soil type and pH value, degree of drainage depending on the species; with low to no tolerance for salt; fruits attract birds and other wildlife; propagate by seed (producing offspring true to type), grafting or stem cuttings.

Drawbacks: Thorny plant so it is hard to prune or work around; seeds are poisonous; not for the superstitious.

Plant Description:
- Small, deciduous, usually thorny trees and shrubs.
- The alternate, simple leaves possess lobed and toothed or deeply lobed margins.
- The usually fetid smelling flowers grow in flat-topped clusters and have 5 sepals, 5 white petals (in rare cases, the petals turn pinkish with age), 5 to many stamens, and 1–5 pistils (which are basally joined).
- The fruits are small pomes (the same fruit type as the apple or pear) with persistent sepals at the tip.

Photos: (opposite top) David Wright; (opposite bottom) Michael Loudon

Crataegus species flowers

Solitary *Crataegus* species tree on Mount Slemish, Ireland

Folklore, Beliefs, Uses and Other Interesting Facts

- In the British Isles, the hawthorn is both auspicious and inauspicious. One will be protected from lightning if one has a hawthorn tree in the garden, shelters under a hawthorn tree during a storm or collect hawthorn on Palm Sunday or Ascension Day and keep it in the house (a Staffordshire variant of the belief). The hawthorn also has healing powers. On the other hand, it is believed to be inhabited by fairies (particularly isolated trees), so it is unsafe to sit under it on special days, such as May Day, Midsummer's Eve or Halloween, for fear of enchantment or abduction. It is unlucky to fell a hawthorn and to bring hawthorn flowers into the home, as doing so spells misfortune or death.
- More recently, the hawthorn is considered as one of the most unlucky of plants in the British Isles. Roy Vickery found that hawthorn flowers were mentioned in 23.5 per cent of 'unlucky' items, with the second most unlucky plant being the lilac (*Syringa* species).
- This belief could have derived from trimethylamine, a chemical shared by both hawthorn and decaying animal flesh. In fact, the smell of hawthorn has been compared with the smell of the Great Plague in London!
- However, in many traditionally Catholic countries, such as Ireland, France and Italy, hawthorn blossoms are believed to be the flower of the Virgin Mary and, therefore, held in high esteem.
- In Ireland, hawthorn is brought into the home on 1 May (May Day) to protect the household from evil.
- Based on feng shui beliefs, the hawthorn is not a plant for the home garden because of its thorns (which are considered poison arrows that emit hostile, inauspicious energy), white flowers (used in mourning or funerals), and its smell of death.
- Some *Crataegus* species are cultivated in Asia, Central America, and various Mediterranean countries for their edible fruit. Some species have fruits with a higher vitamin C content than citrus fruits.
- Many species are cultivated as ornamental shrubs and trees in parks and gardens or in wildlife corridors because their fruit and thorny branches make good nesting sites for birds.

Curcuma longa turmeric • yellow ginger

Scientific Species Name:
Curcuma longa
(= *Amomum curcuma*, *Curcuma domestica*) (Curcuma is from the Arab *al-kurkum*; saffron and *longa* is Latin for long)

Common Species Name:
turmeric, curcuma, Indian saffron, long rooted curcuma, yellow ginger (English); *arrow-root de l'Inde*, *cucurma*, *curcuma long*, *safran des Indes* (French); *Gelbwurzel*, *Gelbwurz*, *Gilber Ingwer*, *Gilbwurzel*, *Indischer Safran*, *Kurkuma* (German); *haldii*, *haldi* (Hindi); *kunyit*, *temu kunyit* (Indonesian, Malay); *huáng jiāng* [黄姜], *yù jīn* [郁金] (Mandarin); *azafrán de la India* (Spanish)

Scientific Family Name:
Zingiberaceae

Common Family Name:
ginger family

Natural Distribution:
Turmeric is a cultigen (cultivated plant which has diverged sufficiently during cultivation from its wild relatives to become a new species) thought to have originated in India.

USDA ≥11 (AUS ≥5)

How to Grow: Ideal temperature range 20–35°C (68–95°F), tropical conditions; moist but not waterlogged soils, pH 5–7.5, but not alkaline conditions; propagated by cutting large or 'mother' rhizomes into pieces, or growing the 'fingers' or 'daughter' rhizomes which branch off the larger ones.

Drawbacks: Leaves may be infected with fungal leaf spot disease; rhizomes may be susceptible to fungal root rot; shoots may be attacked by shoot borer caterpillars, and rhizomes by scale insects.

Plant Description:

- The plant consists of a rhizome from which leafy stems rise to 1 m (3.3 ft) tall.
- Its leaf blades are elliptic and light green in colour.
- The pale yellow flowers are presented in a terminal, cylindrical flowering shoot about 15 cm (5.9 in) long, arising from the rhizome at the base of the leafy shoots, and consisting of pale green to white floral bracts.
- The deep yellow-coloured rhizome has a spicy, fragrant aroma which is well-utilised in Malay and Indian cuisine.

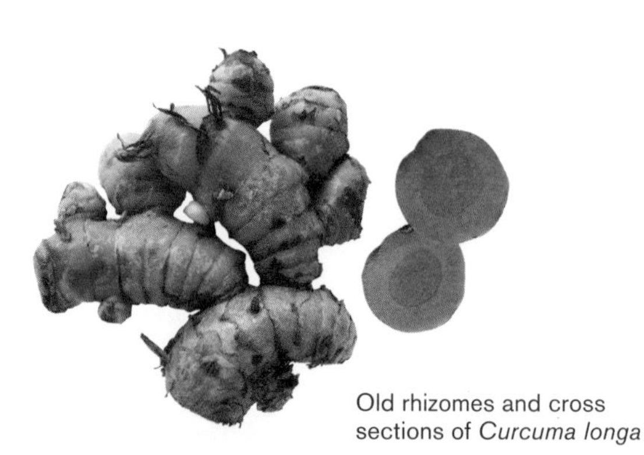

Old rhizomes and cross sections of *Curcuma longa*

Photos: (above and opposite top) Hugh Tan Tiang Wah; (opposite bottom) Jana Skornikova

Leafy shoots constituting the above ground parts of the plant

Curcuma longa infloresence

Folklore, Beliefs, Uses and Other Interesting Facts

- Turmeric is auspicious in Hindu households because yellow is associated with Lord Vishnu, the procreator of the universe. It is an item included in religious rituals, as well as with birth, marriage, death and agriculture.
- Yellow also symbolises fertility, luck and prosperity. Powdered rhizomes are used in many Hindu religious ceremonies. It is used on a bride's face during the wedding day (as with the Malays) to bless her with prosperity and fertility. A deceased wife is also draped in cloth that has been dyed with turmeric for cremation.
- According to feng shui beliefs, the generally green colour of the bushy plant, that does not frequently flower, is auspicious for planting in the south, east or southeast sectors of the garden; neutral for the west, northwest, north sectors; and inauspicious for southwest and northeast.
- The dried rhizome is a culinary spice for curries and previously used as a source of orange or yellow colouring for pharmaceuticals, confectionaries, processed foods, sauces, and dyes for silk, wool and carpets. The fresh rhizome is used for colouring and flavouring rice dishes such as the Malay *nasi kunyit* and the Indian *beriyani*.
- The rhizome is used in Asia as a cosmetic, a traditional cure for stomachache, a tonic and blood purifier, and a cure for the common cold, skin infections and discharge of pus from the eye. Rhizome extracts have pharmaceutical activity against cancer, the HIV virus, inflammation, high cholesterol levels and indigestion. Turmeric is also a fungicide, insecticide and nematicide.
- The young rhizomes and shoots are used as a spicy vegetable and, in India and Nepal, the leaves are added to flour to produce a medicinal bread.

Dendranthema ×grandiflora

chrysanthemum

Scientific Species Name:
Dendranthema ×grandiflora
derived cultivated varieties
(= *Chrysanthemum ×morifolium*)
(Greek *dendron*, tree; *anthemon*, flower, referring to the woody stems that bear the flower clusters; × = multiplication sign, notation for a hybrid; Latin *grandiflorum*, large-flowered, referring to the large flower cluster which resembles a large flower found in these plants; Greek *chrysos*, gold; *anthemon*, flower; Latin *morifolium*, mulberry leaved)

Common Species Name:
chrysanthemum, florist's chrysanthemum, garden mum, garden chrysanthemum, mum (English); *chrysanthème* (French); *Chrysantheme* (German); *daudi*, *guladaudi* (Hindi); *bunga kekwa* (Malay); *jú* [菊], *jú huā* [菊花] (Mandarin); *crisantemo* (Spanish)

Scientific Family Name:
Asteraceae or **Compositae**

Common Family Name:
sunflower family

Natural Distribution:
Possibly a complex hybrid group developed in China from *Dendranthema indica*, which is from East Asia. The chrysanthemum cultivated varieties are bred and selected from this group.

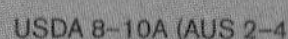
USDA 8–10A (AUS 2–4)

How to Grow: Temperate plant, not recommended for warmer climates; tolerant of mild frost; well-drained, acidic to alkaline, composted sand, loam or clay soils; fairly drought-tolerant after establishment; propagate by dividing the plants or stem cuttings.

Drawbacks: Fairly pest and disease-free; spider mites in hot, dry weather; leaf blight.

Plant Description:

- A perennial herb which grows to 1.5 m (4.9 ft) tall.
- Alternate, stalked leaves, 10–12 cm (3.9–4.7 in) long. The leaves have blades which are dark grey-green, egg-shaped, pinnately-lobed and toothed.
- The flowers are grouped into a head-like cluster or inflorescence (head or capitulum) consisting of disc and ray, or only ray florets (small flowers). Numerous colours and combinations are available, with the exception of blue.

Dendranthema xgrandiflora
cultivar potted plant

 Photos: (above and opposite) Hugh Tan Tiang Wah

Assortment of inflorescences of *Dendranthema ×grandiflora* cultivars

White-flowered cultivar used in a funeral wreath

Top view of inflorescence of *Dendranthema* cultivar

Folklore, Beliefs, Uses and Other Interesting Facts

- Chrysanthemum cut flowers are used in a variety of Chinese religious ceremonies as offerings. Birthdays of deities in Chinese folk religion are often celebrated by laying out a large spread of cooked food and flowers such as chrysanthemum and celosia (*Celosia argentea*). The daily worship of deities such as *guān yīn pú sà* [观音菩萨/觀音菩薩] (Goddess of Mercy, the Bodhisattva of compassion) and family ancestors involves the placement of a vase of chrysanthemum flowers of mixed colours at the altar. Flowers for opening ceremonies for businesses are usually roses (*Rosa* cultivars) or yellow chrysanthemums to symbolise success and prosperity.

- During the 15 days of the Chinese New Year, potted chrysanthemum plants bearing brightly coloured flower heads are purchased to be displayed at the house front for their auspiciousness (good fortune, great prosperity and vitality). The Mandarin name for chrysanthemum, *jú huā*, is a phonetic pun for 'lucky flower', while orange and especially yellow represent gold (prosperity), and red and lilac, blood (vitality). A stout plant with dense foliage is preferred, as the densely arranged leaves are seen as the product of a bountiful harvest and will, similarly, bless the family with plentiful wealth.

- This flower is also associated with death and used in funerals in many countries. Most Chinese believe that receiving these flowers as a gift results in ill-luck because white chrysanthemum flowers are used in funeral wreaths or as offerings to the dead during the festival of *qīng míng* [清明] (15th day after the spring equinox). Similarly, chrysanthemums are used in funerals by the Italians, Japanese, Koreans and Polish. For the people of Malta, having chrysanthemum indoors is believed to bring bad luck.

- The chrysanthemum has been cultivated in China since about AD 200 and is considered one of the 'four gentlemen of flowers', the others being bamboo (tribe Bambuseae, summer), orchid (family Orchidaceae, spring) and the plum (*Prunus* species, winter), each representing a season. Chrysanthemum, the autumn flower, symbolises a life of leisure and retirement and merriness.

- Based on feng shui principles, the yellow chrysanthemum (yellow being the most auspicious colour and symbolic of the earth element) is best planted in the southwest, west, northwest and northeast zones of the garden, and is bad for the north and neutral for the rest. It can also be placed in the centre of the home or at the door front.

- The yellow chrysanthemum is the symbol and crest of the Emperor of Japan who occupies the Chrysanthemum Throne.

- In the language of flowers, the chrysanthemum signifies difficulty or cheerfulness under adversity. A red chrysanthemum says "I love you"; a white chrysanthemum signifies truth; and a yellow chrysanthemum signifies slighted love.

- The chrysanthemum is the birth month flower for November.

Dianthus caryophyllus carnation

Scientific Species Name:
Dianthus caryophyllus
(Greek *dios*, divine, and *anthos*, flower, the divine flower or the flower of Zeus, king of the gods of Greek mythology; Greek *karyophyllon*, which derives from *karya* or *kaura*, walnut, and Greek *phyllon*, a leaf, referring to the aroma of walnut leaves) or less commonly, *Dianthus* ×*allwoodii* (a hybrid of *Dianthus caryophyllus* and *Dianthus plumarius*)

Common Species Name:
border carnation, carnation, clove pink, divine flower (English); *oeillet* (French); *Nelke* (German); *kāng nǎi xīn* [康乃馨], *shè xiāng shí zhú* [麝香石竹] (Mandarin); *clavel* (Spanish)

Scientific Family Name:
Caryophyllaceae

Common Family Name:
carnation family

Natural Distribution:
Possibly of Mediterranean origin

USDA 1–11 (AUS 1–7)

How to Grow: Prefers cooler climates; well-drained, improved, neutral to slightly alkaline soil of medium fertility; apply slow release fertiliser 1–2 times or regularly apply soluble fertiliser during the growing season; propagation is by stem cuttings and seeds.

Drawbacks: Susceptible to bacterial diseases (bacterial wilt, bacterial spot), fungal diseases (*Alternaria*, *Botrytis*, *Fusarium*, *Pythium*, *Rhizoctonia*) and insect attack (aphids, mites, cutworms).

Plant Description:

- Perennial or annual herb up to 30 cm (11.8 in) tall.
- The unstalked, greyish green leaves are opposite with consecutive pairs at right angles to each other. The leaf blades are elongated and covered with a white waxy coating.
- The flowers occur in few-flowered clusters or singly at the stem tips. The sepals are joined together into a cup with the tips as lobes; the 5 narrowly stalked, petals can come in almost all colours except for blue. The petals are solid or edged with different colours, with margins that are toothed or smooth. Stamens are 10 or in most cultivars and become petaloid (to create the double flowers). Styles are 2 or more, topping the ovary.
- The fruit is a capsule which split at the tip to release the numerous seeds.

Dianthus caryophyllus
cultivar flower top view

Photos: (above, opposite middle and bottom) Hugh Tan Tiang Wah; (opposite top) Anna Buxton

Dianthus caryophyllus flowers

Flower at the stem tip

Section of flower

Folklore, Beliefs, Uses and Other Interesting Facts

- The carnation is the birth month flower for January and signifies an expression of strong and pure love. The colour of the carnation flower signifies various qualities: pink, pure love; red, lively/ardent pure love; white, young girl; yellow, disdain.
- In some cultures, the carnation is associated with death and bad luck. In France, the carnation is a symbol of bad luck and death. In Italy, white carnations are used for funeral wreaths.
- The carnation is associated with Mother's Day (Mothering Sunday) in the USA and Canada. A red carnation flower is worn when one's mother is alive and a white one if she has passed on.
- Based on feng shui beliefs, carnations are to be planted in different areas of the garden based on their colour. If pink, magenta or red (fire element), then they are auspicious for the front of the house, the south, southwest and northeast, and inauspicious for the west and northwest, and neutral for the rest. If yellow or orange (earth element), then they are auspicious for the southwest, west and northwest; inauspicious for north; and neutral for the rest. If white (metal element), then they are auspicious for the west, northwest and north; inauspicious for the house front (white is associated with funerals and mourning), east and southeast; and neutral for the rest.
- The carnation is a common commercial cut-flower available all year round at florists. It is also grown for its expensive, strongly fragrant oil that is used in soap, bath and massage oils, candles and perfumes.
- The carnation is the national flower of Slovenia and Monaco; a red carnation the national flower of Spain and a popular national flower of the Czech Republic; and a scarlet carnation, the state flower of Ohio (USA).

Dracaena fragrans cornstalk plant

Scientific Species Name:
Dracaena fragrans
(= *Pleomele fragrans*, *Dracaenaderemensis*)
(Greek *drakaina*, female dragon –the red resin, dragon's blood, is extracted from the dragon tree, *Dracaena draco*; Latin *fragrans*, fragrant, from the strong scent of its flowers)

Common Species Name:
compact dracaena, corn plant, cornstalk plant, fragrant dracaena, green corn plant, striped dracaena (for variegated cultivars) (English); *dragonnier d'Afrique tropicale* (French); *tiě shù* [铁树/鐵樹], *xiāng lóng xuè shù* [香龙血树/香龍血樹], *yì wèi lóng xuè shù* [异味龙血树/异味龍血樹] (Mandarin); *dracena de Africa tropical*, *palmillo* (Spanish)

Scientific Family Name:
Ruscaceae (formerly in the *Agavaceae* or *Dracaenaceae*)

Common Family Name:
butcher's broom family
(formerly in the agave or dragon tree family)

Natural Distribution:
Tropical Africa

USDA 1–11 (AUS 1–7)

How to Grow: Tropical, shade-loving, perennial shrub or tree but can tolerate full sun conditions; moist soil but do not over-water; clayey, sandy or loam, acidic to slightly alkaline soils, preferring much organic material; moderately drought tolerant; not salt tolerant; propagated by stem cuttings.

Drawbacks: Usually low-maintenance and hardy but sensitive to soil moisture content; leaves may turn brown when plant is under- or over-watered; occasional attacks by scale insects and mealy bugs; weekly wiping with a damp cloth or spraying plant outside as large wide leaves get dusty easily; leaf spot disease or root rot; mites, thrips and chewing insects.

Plant Description:

- An unbranched or weakly branching shrub or tree that grows to 6–8 m (19.7–26.2 ft) tall. The stem is woody and unusual for a monocotyledon.
- Its most distinctive feature is the crowded, spirally arranged, stalked leaves at the stem tip. Leaf blades are sword-like, up to 1 m (3.3 ft) by 10 cm (3.3 in).
- The white flowers grow in small clustered shoots that arise from a main shoot. Its flowers are very fragrant, especially at night.
- The fruit is round and fleshy, less than 12 mm (0.5 in) across.

Dracaena fragrans
'Massangeana' leafy shoot

 Photos: (above and opposite top) Hugh Tan Tiang Wah; (opposite bottom) Giam Xingli

Clump of *Dracaena fragrans* trees in a landscaped garden

Tree placed at an office entrance for protection and luck. It is pruned to induce branching.

Folklore, Beliefs, Uses and Other Interesting Facts

- The Chinese in Southeast Asia call the cornstalk plant *tiě shù* in Mandarin, which translates to 'iron tree'. The tree is compared to the strength and protective abilities of the metal as it is believed to ward off evil and spirits. It is also believed that cultivating the *tiě shù* brings good luck. Being inimical to evil spirits and auspicious at the same time, it is not surprising that the introduced cornstalk plant is very popular among the Chinese in Asia.
- In a NASA Clean Air Study, it was found that the cornstalk plant had air-filtering abilities as it could effectively remove formaldehyde. NASA recommended that 15–18 good-sized houseplants (in 15-cm/6-in to 20-cm/8-in diameter containers) be used to remove airborne pollutants in a 167 square-metre (1,800 square-foot) house.
- This species is commercially sold primarily as an indoor foliage ornamental, particularly the more attractive variegated cultivated varieties (e.g., *Dracaena fragrans* 'Lemon Lime' or 'Massangeana', both with yellow-streaked leaf blades). They are also grown as living fences in tropical Africa. Lately, it seems to have fallen out of fashion and has been displaced by the lucky bamboo (*Dracaena sanderana*) which suits smaller gardens and homes better.
- Based on feng shui beliefs, the usually green leafy appearance of the typical green form with solid green leaves suits the wood element sectors of the house, so it can be planted auspiciously in the east, southeast and south of the garden. It is inauspicious for the southwest and northeast, and neutral for the rest. The yellow-streaked leaf cultivars signify the earth element, so it is auspicious to grow the plants in the southwest, west, northwest and northeast of the garden; inauspicious for the north; and neutral for the rest. A plant should not be allowed to grow too tall with a long leafless stem because then it would resemble a 'poison arrow' which is believed to emit inauspicious, hostile energy. To keep the plant short and bushy, prune its top off so more vigorous branching and leafing occurs.

Dracaena sanderana • *Dracaena sanderiana*

lucky bamboo • Chinese water bamboo

Scientific Species Name:
Dracaena sanderana or ***Dracaena sanderiana***
(Greek *drakaina*, female dragon–the red resin, dragon's blood is extracted from the dragon tree, *Dracaena draco*; *sanderana* or *sanderiana*, after Henry Frederick Conrad Sander [1847–1920], British horticulturist famed for his orchid nurseries)

Common Species Name:
Belgian evergreen, Chinese water bamboo, curly bamboo, friendship bamboo, lucky bamboo, Goddess of Mercy plant, ribbon dracaena, ribbon plant, Sander's dracaena, water bamboo (English); *bambou d'eau*, *bambou porte-bonheur*, *canne Chinoise*, *canne de Chine*, *dracaena de Chine* (French); *panaschierter Drachenbaum* (German); *bù bù gāo shēng* [步步高升], *yín yè lóng xuè shù* [银叶龙血树/銀葉龍血樹], *zhuǎn yùn zhú* [转运竹/轉運竹] (Mandarin); *dracena de Camerún* (Spanish)

Scientific Family Name:
Ruscaceae (formerly in the Agavaceae or Dracaenaceae)

Common Family Name:
butcher's broom family (formerly in the agave or dragon tree family)

Natural Distribution:
West–Central Tropical Africa (Cameroon)

USDA ≥11 (AUS ≥5)

How to Grow: A tropical herb or shrub of semi-shaded environments; most soils; can tolerate damp ground or even grow in water; propagate by stem cuttings.

Drawbacks: Can become untidy with growth, so needs pruning; becomes rangy when in overly shady conditions; browning of leaf blade edges and tips caused by excessive sunlight or chlorine in tap water; bacterial leaf spot can afflict; water in vase of the plant may breed mosquito larvae.

Plant Description:

- A perennial, slender herb that grows to about 2 m (6.6 ft) tall.
- Stems and leaves are glabrous and glossy. Leaves, up to 30 cm (11.8 in) long and 4 cm (1.6 in) across are narrowly lanceolate-elliptic and spirally arranged.
- White flowers develop in a cluster at the stem tip, each with 6 shiny white, elongated tepals, and 6 white elongated stamens, and a long white style.

Silver lucky bamboo plant

Golden lucky bamboo plant

 Photos: (above and opposite) Hugh Tan Tiang Wah

Dracaena sanderana cuttings in boat shape

Plants with interwoven stems for sale in a nursery

Flowering shoot at stem tip

Folklore, Beliefs, Uses and Other Interesting Facts

- Lucky bamboo is a misnomer because this plant is not a bamboo, which belongs to the grass family (Poaceae) and tribe Bambuseae. Lucky bamboo is still regarded as an auspicious plant like the bamboos, which the Chinese revere as auspicious because they symbolise resilience to hardship owing to their ability to withstand seasonal changes (especially in winter, when its branches and stems bend but do not break under the weight of the snow). The smaller sized lucky bamboo is a convenient substitute for the rather bulky true bamboo in small homes.
- Different kinds of the solid green (wild type) lucky bamboo are sold, and they are believed to bring luck to different situations. The curly bamboo (with stems trained into spirals) is said to be auspicious for people in a bad patch and thus hoping for a change of luck. Curly bamboo, or *zhuǎn yùn zhú* in Mandarin, is able to 'turn' the bad luck of this person into positive good luck. The various pyramid-tiered arrangements of lucky bamboo, commonly known in the USA as lucky bamboo, or by the Chinese as *bù bù gāo shēng*, are popular as ornamental indoor table-top plants. They are meant to be auspicious for people who are hoping for positive personal and/or career development, signified by the upward pointing shape of the pyramid. Such forms of lucky bamboo are often purchased for Chinese New Year celebrations to induce a change in fortune. There is a cultivated variety with yellow-streaked leaves (golden lucky bamboo) and another with white-streaked leaves (silver lucky bamboo) grown for prosperity.
- The Chinese also believe that lucky bamboo brings good fortune, happiness and prosperity to owners, as well as a harmonious and peaceful life, longevity and health. It also makes them gracious, honourable and virtuous, energizes their love life and creates positive *qì* energy in their environment.
- Based on feng shui beliefs, the usually green, auspicious plants are auspicious if placed near the front door of the house, east, southeast and south sectors; inauspicious for the southwest and northeast; and neutral for the rest.
- This plant is mainly grown commercially as a foliage ornamental pot plant or for floral arrangements.

Epiphyllum oxypetalum queen of the night

Scientific Species Name:
Epiphyllum oxypetalum
(Greek *epi*, upon; Greek *phyllon*, a leaf; referring to the flowers which are borne on flattened, green stems, previously interpreted as leaves; Greek *oxys*, sharp; Botanical Latin *petalum*, petal, referring to the pointed petal tips)

Common Species Name:
Dutchman's pipe, Dutchman's pipe cactus, lady of the night, night blooming cereus, queen of the night (English); *epiphyllum à large feuilles* (French); *bakawali* (Malay); *tán huā* [昙花/曇花] (Mandarin); *reina de la noche* (Spanish)

Scientific Family Name:
Cactaceae

Common Family Name:
cactus family

Natural Distribution:
Mexico to Guatemala, but widely introduced elsewhere

USDA 1–11 (AUS 1–7)

How to Grow: Relatively hardy, tropical cactus; most soils; do not overwater as this is an epiphyte which grows out of the ground on other plants, so intolerant of waterlogging; propagate by woody or green stem cuttings or by seed.

Drawbacks: Can become untidy in appearance and require pruning or tying; flowers bloom infrequently and at night.

Plant Description:

- A sprawling, shrubby, epiphytic, evergreen, perennial, spineless cactus growing to 2–6 m (6.6–19.7 ft) tall.
- Old stems are cylindrical, woody, up to more than 2 m (6.6 ft) long; the younger branches are green, flattened and leaf-like, becoming more flattened towards the tip, 15–100 cm (5.9–39.4 in) long by 5–12 cm (2.0–4.7 in) wide, with scalloped and wavy margins and central ribs.
- The fragrant flowers open at night, are funnel-shaped, and 25–30 cm (9.8–11.8 in) long and 10–27 cm (3.9–10.6 in) across. The sepals are curved towards the tip, light green to dark pink, narrow and elongated. The petals are white, 7–10 cm (2.8–3.9 in) long and 3–4.5 cm (1.2–1.8 in) wide. The numerous stamens have white filaments and cream anthers. The white style is 20–22 cm (7.9–8.7 in) long, bearing 15–20, cream-coloured stigmata at its tip.
- It rarely fruits. When it does, it produces a purplish red, oblong berry about 16 cm (6.3 in) long, similar to that of the dragon fruit or pitaya (*Hylocereus undatus*).
- The many, glossy black seeds are 2–2.5 mm (0.08–0.10 in) long by 1.5 mm (0.06 in) wide.

Photos: (opposite top) Elisa Arteaga; (opposite bottom) Dave and Sarah Thomas

Epiphyllum oxypetalum flowers on a branching stem

Potted plant

Folklore, Beliefs, Uses and Other Interesting Facts

- *Epiphyllum oxypetalum* was first introduced into China in AD 1645. Its beautiful, large, fragrant blooms of the night led to its depiction in ink brush paintings, poems and artistic compositions. Some beliefs also became associated with this ornamental plant. Owing to its slow growth and hardiness, it is reputed for bestowing longevity to the plant's owner.
- The flower blooms about 25 days after an exceptional drop in the temperature. Flowers begin opening at about 9 pm, and complete their opening in about 2 hours. Unfortunately, the flower only lasts for a single night and will wither as the dawn breaks.
- Owing to the rarity of its flowering, most Chinese believe that the flowering of the queen of the night is an auspicious event, indicating that good fortune is to come to the family of the owner. The Chinese saying, *tán huā yī xiàn* [昙花一现/曇花一现], means "the flower of the queen of the night blooms briefly", and is a metaphor for a transient event.
- However, Chinese from the Hakka dialect group regard the flowering of the queen of the night as an inauspicious event because of its rarity.
- Malays of Southeast Asia associate fragrant blooms at night with ghosts or the *pontianak* as this creature is associated with a sickly, sweet scent.
- This species is cultivated as a spectacular ornamental plant in many countries. In China, the flower is frequently eaten in a vegetable soup.

Euphorbia pulcherrima

poinsettia • Christmas plant

Scientific Species Name:
Euphorbia pulcherrima
(After Euphorbus, Greek physician of Juba II, King of Mauretania; Latin *pulcherrima*, pretty, after the attractive floral bracts of this plant)

Common Species Name:
Christmas flower, Christmas plant, Christmas star, flame leaf flower, lobster plant, Mexican flame leaf, painted leaf, poinsettia, winter rose (English); *étoile de Noël*, *poinsettia* (French); *Adventsstern*, *Christstern*, *Poinsettie*, *Weihnachtsstern* (German); *kastuba*, *poinsettia* (Malay); *yì pǐn hóng* [一品红/一品紅] (Mandarin); *flor de Pascua*, *poinsettia* (Spanish)

Scientific Family Name:
Euphorbiaceae

Common Family Name:
spurge family

Natural Distribution:
Mexico

USDA ≥9 (AUS ≥3)

How to Grow: Can tolerate shade, but becomes rangy with excessive shade; keep soil mulched when planted in ground, but when in pot allow to dry out before watering as excess water causes leaf drop; frost kills back plant; do not prune after mid-August to keep the buds which become flowers in winter (northern hemisphere); propagate by stem cuttings in summer.

Drawbacks: Plant is slightly toxic

Plant Description:

- A slender tree or shrub that can grow up to 3 m (9.8 ft) tall, but usually grown as a small pot plant. Plants propagated from rooted cuttings are kept short using growth retardants to produce compact pot plants.
- The alternate, stalked leaves are 8–13 cm (3.1–5.1 in) long and leaf blades are egg-shaped to broadly-elliptic. Its lower leaves are green and sometimes toothed, while the usually smooth margined upper leaves are modified into brightly coloured, petal-like bracts (modified leaves) below the flower clusters at the stem tip. Many cultivated varieties are marked by differences in the colour and shape of the bracts (modified or specialised leaves).
- Flowers are small, usually yellowish.

 Photos: (opposite) Hugh Tan Tiang Wah

Assortment of *Euphorbia pulcherrima* plants with pink, white and variegated red floral bracts

Floral bracts and flowers

Folklore, Beliefs, Uses and Other Interesting Facts

- The poinsettia's brightly coloured floral bracts attract bird pollinators in its natural habitat.
- The association of this plant with Christmas in the northern hemisphere is because it develops flowers when exposed to long night periods, such as in winter. Poinsettias represent 85 per cent of plant sales during Christmas in the United States and 90 per cent of the poinsettias are produced by the USA. All the USA states grow this plant, with California being the biggest producer. Poinsettias are named after Joel Roberts Poinsett (1779–1851), physician, botanist and American minister to Mexico. Poinsett was responsible for introducing poinsettia to the USA and the 12th December, the day he died, is National Poinsettia Day there.
- Based on feng shui principles, the bright red bracts represent fire, so it is auspicious for these plants to be placed in the south or southwest or northeast sectors of the garden; inauspicious in the west, northwest, east and southeast; and neutral for the others. However, because red (representing life blood) is a very auspicious colour, this plant may be placed at the house front for good luck and prosperity.
- In Singapore and Malaysia, this plant is used by the Chinese both for Christmas and the Chinese New Year as red signifies wealth and good luck.

Ficus benjamina **Benjamin fig • weeping laurel**

Scientific Species Name:
Ficus benjamina
(Latin *ficus*, the edible fig [*Ficus carica*]; *benjamina*, from banyan, which is derived from the Sanskrit *banji*, the banyan tree [*Ficus benghalensis* and others])

Common Species Name:
Benjamin fig, Benjamin's fig, Benjamin tree, golden fig, Java fig, Java fig tree, Java tree, tropic laurel, weeping chinese banyan, weeping fig, weeping laurel, (English); Benjamin-Gummibaum, *Birkenfeige* (German); *beringin, jawi jawi, jejawi, kelat sega, waringin* (Malay); *bái róng* [白榕], *chuí yè róng* [垂叶榕/垂葉榕] (Mandarin); *árbol Benjamín, Benjamina, ficus Benyamina, matapalo* (Spanish)

Scientific Family Name:
Moraceae

Common Family Name:
mulberry family

Natural Distribution:
India, South China, Southeast Asia to Australia and Solomon Islands

USDA ≥10b (AUS ≥5)

How to Grow: Tropical strangler or tree; most well-drained soils, acidic to alkaline, but not wet ground; high drought tolerance; moderate aerosol salt tolerance; propagate by seed or stem cuttings.

Drawbacks: Not for the superstitious; frost-sensitive; fruiting individuals through bird dispersers may be the source of seedlings growing on walls, drains, trees, etc.; bird droppings under fruiting trees; windfall fruits from trees; crown throws such dense shade that it excludes grass beneath; thin bark so prone to mechanical damage; drooping branches may need pruning to prevent obstruction; roots of large trees can rip up driveways, roads, drains, etc.; grows so large, it will take over a small garden!

Plant Description:

- An evergreen tree which usually starts as an epiphyte on another tree then strangles it with its aerial roots or an independent tree which can grow up to a height of 35 m (114.8 ft) tall and wide. It has characteristically drooping branches with long aerial roots emerging below. White latex oozes from all cut surfaces.
- The stalked leaves are alternate and the leaf blades are 2–14 cm (0.8–5.5 in) by 1.5–8 cm (0.6–3.1 in), elliptic, oblong to obovate, thinly leathery, shiny dark green above, with wavy margins and a prolonged drip tip.
- The tiny flowers grow inside a round, 8.5–12.5 mm (0.3–0.5 in) wide structure called a syconium, which ripens yellow to orange to dark red or pink to purple.
- The tiny fruits, each with one seed, develop from the tiny flowers in the syconium which, when ripened, attracts birds which eat the fruits and disperse the seeds in their droppings.

Ripening syconia on leafy twigs

 Photos: (above and opposite) Hugh Tan Tiang Wah

Ficus benjamina tree with drooping branches strangling its host tree

Topiary plants

Plants with pleated stems

Folklore, Beliefs, Uses and Other Interesting Facts

- Large trees, especially fig trees, are thought by the Malays of Southeast Asia to house *datuks* (good, benign or evil spirits), ghosts or the *pontianak* (a woman who died at childbirth to become undead, a female vampire who terrorises the living). The tree's size, dark shady crown and aerial roots hanging from the branches lend to this effect, especially since the *pontianak* is thought to swing from the aerial roots. Tree cutters in Singapore and Malaysia will offer prayers before felling large Benjamin fig trees to appease the spirit believed to reside within. If the tree is large enough, tree cutters will often refuse to fell it.
- Some Chinese associate large fig trees with good spirits and will ask it to adopt a sick child to improve the child's health.
- An old Malay custom is to plant two fig trees in front of a royal home.
- Based on feng shui beliefs, the Benjamin fig is not auspicious to plant because its drooping branches signify sorrow, and the dense crown blocks the flow of *qì* (energy or life force) allowing stale *qì* to accumulate.
- The Benjamin fig is a popular ornamental species which is grown as a bonsai plant, hedge, standard, potted with the stems pleated (as in *Pachira aquatica*), as a potted small tree, topiary and as a tree.
- The Benjamin fig is often confused with the Malayan banyan (*Ficus microcarpa*) and can be distinguished as below:

	Ficus benjamina	*Ficus microcarpa*
Leaf blade	shiny dark green above	matte mid-green above
Leaf blade margin	wavy (undulating)	flat
Leaf blade tip	prolonged into a drip tip	rounded to somewhat pointed
Branches	drooping	horizontal to ascending
Aerial roots	few from branches, usually string-like, not forming a dense curtain beneath the crown	many from branches, usually pillar-like, forming a dense curtain beneath the crown

Ficus carica common fig • cultivated fig

Scientific Species Name:
Ficus carica
(Latin *ficus*, this species; Latin *carica*, Greek *karike*, this species)

Common Species Name:
common fig, cultivated fig, edible fig, fig, wild fig (English); *figuier commun* (French); *echte Feige*, (German); *anjir* (Hindi); *anjir* (Malay); *wú huā guǒ* [无花果/無花果] (Mandarin); *higo*, *higuera común* (Spanish)

Scientific Family Name:
Moraceae

Common Family Name:
mulberry family

Natural Distribution:
Probably Southwest Asia but spread early to the rest of the Mediterranean

USDA 8–10 (AUS 2–4)

How to Grow: Hardy Mediterranean shrub or tree; with winter protection, can grow to zone 5, and when completely dormant can tolerate temperatures to −9°C (15.8°F); well-drained, organic soils of pH 6.0–6.5; mulch to conserve soil moisture; propagate by stem cuttings, root suckers, ground or air-layering, grafting, seed or tissue culture.

Drawbacks: The latex of the unripe fruits and other plant parts may be severely irritating to human skin if not cleaned off quickly; hot-dry spells can cause fruit drop; soil nematodes can be reduced by a heavy mulch; insect pests include leaf and shoot-eating beetles, stem-borers, fig borers, scale insects; fungal diseases include leaf rust, leaf spot and fruit smut; trees infected with fig mosaic virus disease must be destroyed.

Plant Description:

- A latex-producing, deciduous shrub or tree to about 9 m (29.5 ft) tall.
- The simple, alternate leaves have leaf blades with 3–7 lobes, growing up to 25 cm (9.8 in) long and across, and softly hairy beneath.
- The egg-, top- or pear-shaped 'fruits' are actually syconia (singular syconium; confusingly also called 'figs') and develop in the leaf axils. They are 2.5–10 cm (1–3.9 in) long. The syconium consists of a hollow, fleshy structure, lined on the inside by tiny flowers with a tiny hole at its tip. This species consists of plants that produce male and female flowers in the syconium that develop into inedible syconia (caprifigs) containing wasps (its pollinator species), and plants that produce only female flowers in the syconia that develop into edible syconia containing seeds. A third kind of plant, derived by mutation, produces edible syconia but without pollination or fertilisation, and, hence, is seedless.
- Seeds are large to minute and number from 30 to 1,600 per syconium.

Ficus carica seeds

Photos: (above) Steve Hurst @ USDA-NRCS PLANTS Database; (opposite top and middle) David Karp, ARS, USDA; (opposite bottom) Phillip Merritt

Ficus carica 'Panachee' with striped syconia

Ficus carica 'Black Mission' ripe syconia in a crate in California

Ficus carica shrubs in cultivation

Folklore, Beliefs, Uses and Other Interesting Facts

- In the book of Genesis in the Bible, Adam and Eve covered their genitalia with fig leaves upon realising that they were naked after eating from the Tree of the Knowledge of Good and Evil: "And the eyes of them both were opened, and they knew that they were naked; and they sewed fig leaves together, and made themselves aprons." (Genesis 3:7, King James Version).
- Similarly, many paintings done in Europe depict fig leaves protecting the modesty of otherwise nude men and women. This is most likely in reference to the origin story of Adam and Eve in the Bible. The fig leaf has come to be seen as a covering for human modesty and, at the same time, a symbol of sexuality.
- In Deuteronomy 8:7–8 (King James Version), the people of Israel were promised: "....a good land... A land of wheat, and barley, and vines, and fig trees, and pomegranates; a land of olive oil, and honey". These plants thus have a special place in Judaism as the seven plants of the Bible. The common fig is also one of the most commonly cultivated plants of the Bible lands. The word 'fig' or 'figs' is mentioned at least 44 times in the Bible.
- Blessing is often signified by the grape and the fig together: "And Judah and Israel dwelt safely, every man under his vine and under his fig tree, from Dan even to Beersheba, all the days of Solomon." (1 Kings 4:25, King James Version). Or "But they shall sit every man under his vine and under his fig tree; and none shall make them afraid: for the mouth of the LORD of hosts hath spoken it." (Micah 4:4, King James Version).
- The fig is also mentioned in the Koran, one of the less than 20 plant species mentioned: "I swear by the fig and the olive."(95:1, translation by M. H. Shakir).
- In the language of flowers, the fig signifies longevity or prolific argument.
- Although it is generally accepted that *Ficus carica* was domesticated about 6,500 years ago, a 2006 report of a discovery in the Jordan Valley suggests that it may have been domesticated up to 11,400 years ago. This would make the fig the first domesticated plant of the Neolithic Revolution (the first agricultural revolution), predating, by about a millennium, the domestication of cereal.

Ficus microcarpa laurel fig • Malayan banyan

Scientific Species Name:
Ficus microcarpa
(Latin *ficus,* the edible fig [*Ficus carica*]; Greek *mikros*, small; Greek *karpos*, fruit, referring to the small syconia borne on the twigs)

Common Species Name:
Chinese banyan, curtain fig, Indian laurel fig, laurel fig, Malay banyan, Malayan banyan (English); *arbre de l'Intendance* (French); *Chinesische Feige*, *Indischer Lorbeer*, *Lorbeerfeige* (German); *chilkan*, *kamarup* (Hindi); *jejawi* (Malay); *róng shù* [榕树/榕樹] (Mandarin); *laurel de Indias* (Spanish)

Scientific Family Name:
Moraceae

Common Family Name:
mulberry family

Natural Distribution:
Sri Lanka to India, China, Ryukyu Islands, Bonin Islands, Southeast Asia, Christmas Island, Australia, Solomon Islands, Palau and Truk Islands and Cocos Island

USDA ≥10b (AUS ≥5)

How to Grow: Tropical strangler or tree; most soils, acidic to alkaline, tolerant of wet ground; moderate drought tolerance; high aerosol salt tolerance; propagate by seed or stem cuttings.

Drawbacks: Not for the superstitious; frost-sensitive; fruiting individuals (through bird dispersers) may be the source of seedlings growing on walls, drains, trees; bird droppings under fruiting trees; windfall fruits create a mess under the trees; crown throws such dense shade that it excludes grass beneath; roots of large trees can rip up driveways, roads, drains; grows so large that it will take over a whole small garden.

Plant Description:

- A tree growing to 30 m (98.4 ft) tall and wide, with a flat or round crown, or a shrub. Branches are horizontal to ascending, with a dense curtain of aerial roots hanging down, or some become pillar-like to support the widespread branches. White latex oozes from all cut surfaces.
- The stalked, alternate leaves have slightly fleshy, leathery, elliptic to broadly elliptic leaf blades 2–14 cm (0.8–5.5 in) by 1–8 cm (0.4–3.1 in), with matte, mid-green upper sides, with flat margins and rounded to pointed tips.
- The tiny flowers grow inside a round, 5–15 mm (0.2–0.6 in) wide structure called a syconium.
- The tiny fruits, each with 1 seed, develop from the tiny flowers in the syconium which, when ripened, is a bird-attracting pink to dark purple.

 Photos: (opposite top) Hugh Tan Tiang Wah; (opposite middle and bottom) Angie Ng-Chua

Ficus microcarpa tree

Ripe syconia on leafy branches

Ripe syconia on leafy twigs

Folklore, Beliefs, Uses and Other Interesting Facts

- *Ficus microcarpa* is often confused with *Ficus benjamina* because of their superficial similarity. As is the case with the latter, *Ficus microcarpa* is believed to house *datuks* (good, benign or evil spirits), ghosts or the *pontianak* (a woman who died at childbirth to become undead, a female vampire, who terrorises the living). Like *Ficus benjamina*, it also has a dark shady crown with aerial roots hanging from the branches. Together with its large size, the effect becomes most intimidating especially since the *pontianak* is thought to swing from the roots. Tree cutters in Singapore and Malaysia will offer prayers before felling these trees to appease the spirit believed to reside within, and will often refuse to fell it if it is large enough.
- Some Chinese associate large fig trees like those of Malayan banyan with good spirits and can ask it to adopt a sick child to improve the child's health.
- In Singapore, a Chinese temple has a Malayan banyan tree growing in its compound with an altar set within its pillar-like aerial roots.
- This is a popular ornamental species which is grown as a bonsai plant, hedge (especially a yellow-leafed cultivated variety), standard, as a small potted tree, topiary and as a tree.
- It is often confused with the Benjamin fig (*Ficus benjamina*) and can be distinguished as below:

	Ficus microcarpa	*Ficus benjamina*
Leaf blade	matte mid-green above	shiny dark green above
Leaf blade margin	flat	wavy (undulating)
Leaf blade tip	rounded to somewhat pointed	prolonged into a drip tip
Branches	horizontal to ascending	drooping
Aerial roots	many from branches, usually pillar-like, forming a dense curtain beneath the crown	few from branches, usually string-like, not forming a dense curtain beneath the crown

Ficus religiosa bodhi tree • sacred fig tree

Scientific Species Name:
Ficus religiosa
(Latin *ficus*, the edible fig, *Ficus carica*; Latin *religiosa*, utilised for religious ceremonies—sacred)

Common Species Name:
bo tree, bodhi tree, Indian fig tree, peepul tree, pipal tree, pippala, pipul, po tree, sacred fig, sacred fig tree (English); *arbre bo de*, *figuier de l'inde*, *figuier des banians*, *figuier des pagodes*, *figuier Indien*, *figuier sacré*, *figuier sacré de bodh-gaya* (French); *Bobaum*, *Baum*, *Bodhi-Baum*, *Heiliger Feigenbaum*, *Pappelfeige*, *Pepul-Baum*, *Pepulbaum der Inder* (German); *pipal*, *pipali*, *pipli* (Hindi); *bodi* (Malay); *sī wéi shù* [思维树], *pú tí shù* [菩提树/菩提樹] (Mandarin); *arbol sagrado de la India*, *higuera de las pagodas*, *higuera religiosa de la India*, *higuera sagrada de los Budistas* (Spanish)

Scientific Family Name:
Moraceae

Common Family Name:
mulberry family

Natural Distribution:
India to Southeast Asia

USDA ≥11 (AUS ≥5)

How to Grow: An epiphytic, fast-growing tree which can also grow in the ground; tolerant of drought; not tolerant of frost as it is a tropical tree; propagate by stem cuttings or seed which germinate easily.

Drawbacks: Generally pest- and disease-free; tree can grow over buildings, engulfing them within its roots, or on other trees, ultimately splitting them; seedlings will grow (from seeds dispersed by birds) up cracks in walls, drains or roofs.

Plant Description:

- A more or less deciduous and usually epiphytic (when wild) tree that can grow up to 25 m (82 ft) tall. It has a short trunk of usually not more than 2 m (6.6 ft) but a spreading crown and long, fast-growing branches. The crown is irregularly shaped and yellowish.
- The alternate, long-stalked leaves have thinly leathery, heart-shaped, 7.5–18 cm (3–7.1 in) by 5.5–13 cm (2.2–5.1 in) leaf blades with long, narrow tips.
- The tiny flowers are found inside the cavity of a round, 12 mm (0.5 in) wide syconium (plural of 'syconia'). The flowers form a highly modified inflorescence or arrangement of flowers.
- After the flowers are fertilised, tiny fruits develop in the syconia, which turns greenish yellow, then deep purple when ripe.

 Photos: (opposite) Hugh Tan Tiang Wah

Ficus religiosa tree at the Sri Lankarayama temple in Singapore

Fruits and leaves

Fallen leaves, fruits and tiny seeds

Folklore, Beliefs, Uses and Other Interesting Facts

- The bodhi tree is one of the most ancient and venerated plants, and is sacred to Buddhists and Hindus.
- The bodhi tree is the tree under which Siddharta Gautama (Buddha) achieved enlightenment and went on to spread the Buddhist way. A sprig of that tree was brought from India in 288 BC, was rooted and became the Anuradhapura in Sri Lanka. The presence of the tree in temples and shrines represents Buddha himself; thus respect is always accorded to the bodhi tree by worshippers.
- Hindus believe that the bodhi tree is associated with many deities, the most important one being Vishnu, the creator of the universe. In India, the bodhi tree is regarded as a symbol of the male, whereas the nimtree (*Azadirachta indica*) represents the female. These trees are grown together in villages within a surrounding platform. Intertwined or coiled snake stones, symbolising fertility, are placed on the platform. The sexes represented by the trees are reversed in Punjab and Rajasthan where the bodhi tree is the female instead.
- Owing to its importance to both Buddhists and Hindus, the bodhi tree, especially a mature one, is rarely felled. Many recognise its religious and mythical importance, and most land owners and woodcutters alike are reluctant to perform such a disrespectful act.
- Other than being cultivated for its sacred and auspicious value, its leaves are used for miniature paintings and its fibres were formerly used for paper-making in Myanmar.

Fortunella japonica • *Fortunella margarita*

golden orange • kumquat

Scientific Species Name:
Fortunella japonica or
Fortunella margarita
(*Fortunella* named after Robert Fortune [1812–1880], Scottish collector of plants in China and horticulturist who smuggled out the tea plant [*Camellia sinensis*] to the West; Latin *japonica*, Japanese; Latin *margarita*, a pearl)

Common Species Name (general):
cumquat, cumquot, golden orange, kinkan, kumquat (English); *jīn jú* [金橘] (Mandarin)

Common Species Name (*Fortunella japonica*):
round kumquat, marumi kumquat (English); *kumquat rond, kumquat à fruits ronds* (French); *Rundkumquat* (German); *jīn gān* [金柑], *jīn jú* [金橘], *shān jú* [山橘] (Mandarin); *kumquat redondo, naranjita Japonesa* (Spanish)

Common Species Name (*Fortunella margarita*):
bullet kumquat, oval kumquat, nagami kumquat, pearl lemon, spicy-peeled kumquat (English); *kumquat à chair acide, kumquat ovale, kumquat à fruits oblongs* (French); *Chinesische Kumquat, Oval Kumquat, Ovale Kumquat, Ovaler Kumquat* (German); *jīn jú* [金橘] (Mandarin); *naranjita China* (Spanish)

Scientific Family Name:
Rutaceae

Common Family Name:
citrus family

Natural Distribution:
Fortunella japonica is probably from South China; *Fortunella margarita* is possibly from Southeast China.

USDA ≥11 (AUS ≥5)

How to Grow: Preferring a cool subtropical or warm temperate climate during the growing season but 26–37°C (79–99°F) for optimal growth; frost tolerant; intolerant of drought and flooding; well-drained, rich soil; propagated by seed, air-layering or grafting onto hardy rootstocks.

Drawbacks: Relatively trouble free; fruits are generally not eaten by squirrels or birds; scab, green scurf, greasy spot, anthracnose, fruit rot and stem-end rot diseases; leaf miners, caterpillars, tree borers and, for potted plants, mealybugs.

Plant Description:

- A bush tree growing to 2.4–4.5 m (7.9–14.8 ft) tall but usually grown as a dwarf plant in a pot. Twigs are green with few or no thorns.
- The stalked, alternate leaves have leaf blades that are egg- to lance-shaped with visible tiny glandular spots and are 3–8 cm (1.2–3.1 in) long. The leaves are fragrant when crushed.
- The sweetly fragrant, orange blossom-like flowers are arranged singly or up to fours in clusters at the leaf axils. There are 5 white petals and 5 stamens.
- The fruits are characteristically orange-coloured with numerous oil glands on the thick peel. The oval kumquat, *Fortunella margarita*, bears oval or oblong fruits. *Fortunella japonica*, the round kumquat, bears round fruits. There are 3–6 segments, which contain sour to slightly sour juice.
- There may be a few to no seeds per fruit.

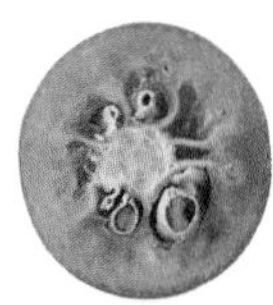

Fortunella margarita fruits in various views

 Photos: (above and opposite) Hugh Tan Tiang Wah

Fortunella margarita fruits on a leafy twig

Fortunella margarita potted plant

Folklore, Beliefs, Uses and Other Interesting Facts

- The Chinese believe that the kumquat is an auspicious plant and a must-have for the Chinese New Year festivities. Because of its golden-orange coloured fruits, the kumquat plant is called *jīn gān* or *jīn jú* in Mandarin, which means 'golden orange'. The numerous fruits borne on the kumquat plant, like an abundance of gold coins, symbolise abundance and wealth. Hence, during the Chinese New Year period, potted fruiting kumquat plants are imported from China where they are commercially cultivated. Buyers pick the plant with the most fruits to maximise the wealth and luck the plant is expected to bring for the coming lunar year. Two potted plants are placed on either side of the front door of the house. If plants are to be ground cultivated in the garden, then they should be within view of the front door, or if the garden is in the back of the house, then they should be planted within view of the back door. The fruits and leaves are also used as decorations during this period. During the Vietnamese New Year, potted kumquat plants are also used for decoration.
- Preserved kumquats are used in a traditional Chinese recipe to cure sore throats and dry coughs. The fruits are placed in a jar of coarse/rock salt for a few months to be dehydrated. These salted, dehydrated fruits can be cut into small pieces and sucked one piece to ease a sore throat or cough. A drink made by mixing pieces of preserved kumquat and honey with hot water can also be administered. The Vietnamese also use the same cure.
- The fruits can be eaten raw, with the peel, in chutneys, marmalades, jellies or in syrup, or candied.
- These plants are popularly grown as ornamentals in the USA in hedges or home gardens.
- A liqueur is prepared from the fruit in Australia.

Four Leaved Clover

The four leaved clover, or shamrock, is an icon of good luck. The shamrock usually has three leaflets per leaf, so one which develops four leaflets is considered lucky because of its rarity. As to which species is the true shamrock is still being debated. The four species listed below are some of the prime candidates and are species of *Medicago* (medicks or burweeds), *Oxalis* (sorrels) and *Trifolium* (clovers).

Scientific Species Name:
Medicago lupulina
(Greek *medike*, medick, the name for a crop plant introduced from Media; Latin *lupulina*, hop-like, referring to the similarity of the inflorescence of this species to that of the hop [Humulus lupulus], which is used in beer brewing).

Common Species Name:
black medick, black trefoil, hop clover, nonesuch, trefoil, yellow trefoil (English); *lupuline*, *luzerne lupuline* [Switzerland], *minette*, *minette dorée* (French); *Gelbklee*, *Hopfenklee* (German); *tiān lán mù xù* [天籃苜蓿/天籃苜蓿] (Mandarin); *alfalfa lupulina*, *carretilla* [Mexico], *carretón*, *fenarola-menuda* [Catalan], *herba de la desfeta* [Catalan], *lupulina*, *melgó menut* [Catalan], *meligón*, *mielga*, *mielga azafranada*, *mielga negra*, *trèvol* [Catalan] (Spanish)

Scientific Family Name:
Fabaceae or **Leguminosae**
(formerly in the *Papilionaceae*)

Common Family Name:
bean family

Natural Distribution:
Europe and temperate Asia but has become naturalised in areas of the tropics and subtropics

USDA 3–10 (AUS 1–4)

How to Grow: Most soil types, but prefers calcareous, well-drained, dry, fine-textures, slightly acid to neutral soils; propagate by seed.

Drawbacks: A major weed of turf areas.

Plant Description:

- A hairy, procumbent or ascending, annual or short-lived, perennial herb 10–80 cm (3.9–31.5 in) long.
- The hairy, alternate, stalked leaves have 3 leaflets, each with an oval to rhombic blade, weakly toothed along the margins towards the tip.
- The axillary, stalked inflorescence bears 10–50 compactly arranged, yellow-petalled flowers.
- The fruit is a kidney-shaped, hairy legume that is black when ripe, containing 1 seed.

Medicago lupulina fully developed and young inflorescences

Photos: (above) Doug Waylett; (opposite) Hugh Tan Tiang Wah

Four Leaved Clover

Scientific Species Name:
Oxalis species or
Oxalis acetosella
in particular (Greek *oxys*, acidic or sour; *acetosella*, is an early name for sorrel (*Rumex* species) or other acidic-leaved plants; Latin *acetum*, vinegar)

Common Names for *Oxalis* species:
sorrel, shamrock (English); *oseille* (French); *Sauerklee, Sauerampfer* (German); *zuò jiāng cǎo* [酢浆草] (Mandarin); *oxalida* (Spanish)

Common Names for *Oxalis acetosella*:
common wood sorrel, cuckoo bread, European wood-sorrel [USA], green snob [USA], Irish shamrock, shamrock [UK], sleeping beauty [USA], white wood sorrel, Whitsun flower [USA], wood sorrel (English); *alleluia, oseille, oseille des bois, oxalide des bois, oxalide petite oseille, oxalis des bois, oxalis petite-oseille, pain de coucou, petite oseille, petite oxalide, surelle, surelle pain-de-coucou, surelle petite oxalide* [Switzerland] (French); *Echter Sauerklee, Gemeiner Wald-Sauerklee, Hainklee* [Switzerland], *Hasenklee* [Switzerland], *Kuckucksklee* [Switzerland], *Sauerklee, Waldklee, Waldsauerklee* (German); *acederilla, acetosilla, agreta, aleluya, garfala, pan de cuclillo, platanito, trébol acedo, trébol ácido, trébol amargo, trébol ácido, vinagrera blanca* (Spanish)

Scientific Family Name:
Oxalidaceae

Common Family Name:
wood sorrel family

Natural Distribution:
***Oxalis* species:** Cosmopolitan (about 700 species);
***Oxalis acetosella*:** Temperate Eurasia

USDA 3–10 (AUS 1–4)

How to Grow: This only applies to *Oxalis acetosella*, so for other species, please check other references; constantly moist soil which should not dry out between watering; propagate by plant division or seed.

Drawbacks: May become weedy; slightly toxic because of the oxalic acid in its tissues which is also responsible for its sour taste, causing kidney damage when eaten in significant quantities by livestock.

Plant Description:

- This applies to *Oxalis acetosella*.
- A perennial herb with a creeping stem.
- The alternate, stalked leaves have three leaflets with oval to heart-shaped, blades folded lengthwise.
- The axillary, slender-stalked flowers have usually pink-streaked petals.
- The fruit is a dehiscent capsule, which splits to release the tiny seeds which eject by their arils (seed outgrowths) turning inside out.

Oxalis corniculata upper and lower leaf views (top) and fruiting and flowering plant (bottom)

Four Leaved Clover

Scientific Species Name:
Trifolium dubium
(Latin *trifolium*, clover; Latin *tri*, three; Latin *folium*, leaf, referring to the leaf which has three leaflets; Latin *dubium*, doubtful, as in not conforming to pattern)

Common Species Name:
lesser trefoil, lesser yellow trefoil, little hop clover, low hop clover, red suckling clover, small hop clover, shamrock, suckling clover, yellow clover, yellow suckling clover (English); *trèfle douteux*, *trèfle étalé*, *trèfle filiforme* (French); *Faden Klee*, *Gelber Wiesen-Klee*, *Zweifelhafter Klee* [Switzerland], *Zwerg-Klee* (German); *huáng shū cǎo* [黄菽草] [Taiwan], *xiàn sān yè cǎo* [线三叶草/綫三葉草], *xiàn yè sān yè cǎo* [线叶三叶草/綫葉三葉草] (Mandarin); *trébol amarillo*, *trébol filiforme* (Spanish)

Scientific Family Name:
Fabaceae or **Leguminosae**
(formerly in the *Papilionaceae*)

Common Family Name:
bean family

Natural Distribution:
Europe, extending to the Caucasus. Has spread to cool and warm temperate grasslands outside its natural distribution.

Trifolium dubium flowering plants

USDA 3–10 (AUS 1–4)

How to Grow: Usually a wild weed in its native range or country of introduction, so not often cultivated; some frost tolerance; low to medium drought tolerance; most types of well-drained to wet soils; propagation by seed.

Drawbacks: Spreads invasively through its creeping stems; can become weedy.

Plant Description:

- A semi-erect to prostrate, annual herb, 20–60 cm (7.9–23.6 in) long.
- The thin, wiry, occasionally downy haired stems are basally branching.
- The hairless to slightly hairy, alternate, stalked leaves have 3 leaflets, each with a narrowly elliptic or obovate, grey-green blade that is broadest at the tip, and the terminal leaflet is stalked.
- The axillary, stalked inflorescence bears 12–30, compactly arranged, lemon yellow-petalled flowers.
- The fruit is a legume which contains a single, yellow to olive, oval seed. There are approximately 2 million seeds per kg (2.2 pounds).

 Photos: (above) Rita van Deemter; (opposite) Maria Porta

Four Leaved Clover

Scientific Species Name:
Trifolium repens
(Latin *trifolium*, clover; Latin *tri*, three; Latin *folium*, leaf, referring to the leaf which has three leaflets; Latin *repens*, creeping)

Common Names for *Oxalis species*:
Dutch clover, honeysuckle clover, ladino clover, shamrock, white clover, white Dutch clover (English); trèfle blanc, trèfle rampant (French); *Lämmer-Klee*, *Weissklee*, *Weiss-Klee*, *Kriechender Klee*, *Kriechender Weiss-Klee* (German); *semanggi landa* (Malay); *bái sān yè cǎo* [白三叶草/白三葉草], *bái chē zhóu cǎo* [白车轴草/白車軸草], *bái huā sān yè cǎo* [白花三叶草/白花三葉草], *shū cǎo* [菽草] (Mandarin); *trébol blanco*, *trébol amargo*, *trébol rastrero*, *trébol de Holanda* (Spanish)

Scientific Family Name:
Fabaceae or **Leguminosae** (formerly in the *Papilionaceae*)

Common Family Name:
bean family

Natural Distribution:
Probably originated in the Mediterranean region, but is indigenous to the whole of Europe, Central Asia (west of Lake Baikal) and to Morocco and Tunisia.

USDA 3–10 (AUS 1–4)

How to Grow: Temperate to subtropical plant; optimal growth at 20–25°C (68–77°F); some frost tolerance, depending on cultivar; low drought tolerance; well-drained, moist, fertile soils with pH 5.8–6.0 on mineral soils and 5.5–5.8 on peaty soils; vulnerable to high levels of available manganese and aluminium; propagation by seed.

Drawbacks: Can become weedy; fungal diseases include those of the creeping stems and roots (*Pythium middletonii*), produce rust pustules on leaves (*Uromyces* species), clover rot (*Sclerotinia trifoliorum*), etc.; viral diseases include alfalfa mosaic, bean yellow mosaic, clover yellow vein, peanut stunt and white clover mosaic; pests include aphids, slugs, snails, weevils, and other herbivorous insects.

Plant Description:

- A creeping, perennial herb.
- The stems are hairless or nearly hairless.
- The hairless, alternate, long stalked leaves have 3 leaflets, each with an elliptic to obovate to heart-shaped blade with slightly toothed margins, with white or light green markings. Leaf stalk length and leaflet size vary with the cultivar.
- The axillary, stalked inflorescence bears up to 40 compactly arranged, white or light pink-petalled flowers.
- The fruit is a narrowly oblong, 4–5 mm (0.16–0.20 in) long legume that bears 3–4, pale yellow (rarely reddish brown), oval to kidney-shaped seeds. There are between 1.45 million and 1.67 million seeds per kg (2.2 pounds).

Trifolium repens inflorescence

Oxalis species flowers

Trifolium repens flowering plants

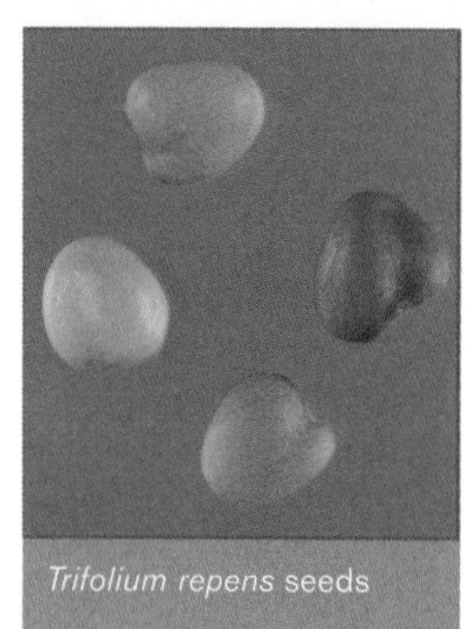
Trifolium repens seeds

Folklore, Beliefs, Uses and Other Interesting Facts

- Shamrock is the anglicised version of the Gaelic (Irish) *seamróg* or *seamair óg* for young clover. Shamrock typically has trifoliate leaves but a quadrifoliate leaf (with four leaflets), probably derived from an error in development of the leaf, are much rarer and believed to bring luck to people who can find them. Hence the term, 'lucky four-leaved clover,' which actually refers to four leaflets of one leaf.
- According to Irish legend, the shamrock was looked upon as a sacred plant because of its trifoliate leaves, three being a mystical number in the Celtic religion. In the British Isles, clover is believed to be an anti-witch plant that protects animals and humans from evil spells and the tricks of fairies, and brings good luck to those who wear it in their hats or buttonholes, or keep it in their homes. Young couples who dream of clover would regard this as a good omen for a joyful and prosperous marriage. While the trifoliate clover has these auspicious qualities, the rarer quadrifoliate kind (four-leaved clover) was especially powerful as it protected the wearer from evil spells and enabled one to detect evil spirits, fairies and witches. Hiding one quadrifoliate leaf in the cow house or dairy protected the milk supply or butter from being harmed by witches.
- Shamrock is worn on the Feast Day of St. Patrick, celebrated on 17 March, to associate with the Wales-born Saint Patrick, the bishop who spread the Christian message in Ireland. It is commonly believed that in the 5th century AD, Saint Patrick used the trifoliate shamrock to explain the concept of the Holy Trinity (God the Father, the Son and the Holy Spirit) to the pagan Irish. Although there is no evidence from old Irish manuscripts that St. Patrick used the shamrock as such, there has been a tradition of wearing shamrock on Saint Patrick's Day since the early 18th century.

Photos: (this page, top to bottom) Scott Bauer, ARS, USDA; Maria Porta; Steve Hurst @ USDA-NRCS PLANTS Database; (opposite page, top to bottom) R R Smith, ARS, USDA; Robert H Mohlenbrock @ USDA-NRCS PLANTS Database; Patrick Standish

Trifolium pratense flowering plants

Medicago lupulina flowering plants

Trifolium repens 'Purpurascens Quadrifolium'

- Shamrock is grown commercially in Ireland to meet domestic and international demands for Saint Patrick's Day festivities. It is germinated from seed and cultivated in the open ground, pots or in a special water-retaining gel.
- The shamrock is an unofficial emblem of Ireland, the national emblem being the harp. The trifoliate shamrock leaf is commonly used to represent anything Irish—from St. Patrick's Day to Irish pubs and restaurants.
- Its strong association with Irish identity began in the 19th century when the shamrock became a symbol of rebellion against the English rule.
- The botanical identity of the shamrock is by no means a certainty, as it is based on artwork, and because there are many species with trifoliate leaves in Ireland. In a survey carried out on Irish people by Charles Nelson for his book Shamrock: Botany and History of an Irish Myth, most Irish identified the shamrock as *Trifolium dubium* (Gaelic *seamair bhuí*) followed by *Trifolium repens* (*seamair bhán*), then *Medicago lupulina*, then *Oxalis acetosella* (*seamóg*), and *Trifolium pratense* (red clover, *seamair dhearg*).
- The cultivated variety, *Trifolium repens* 'Purpurascens Quadrifolium' has all its leaves with four maroon-red and green margined leaflets. Up to 14 leaflets have been found in wild populations of *Trifolium repens* and *Trifolium pratense*. With such plants, one would be lucky everyday!

Fuchsia species & hybrids

fuschia • lady's eardrops

Scientific Species Name:
***Fuschia* species and hybrids**
(Latin *Fuchsia*, after German physician and botanist Leonhart Fuchs [1501–1566], one of the founding fathers of botany)

Common Species Name:
fuchsia, lady's eardrops (English); *fuchsia* (French); *Fuchsie* (German); *dào guà jīn zhōng* [倒挂金钟/倒掛金鐘] (Mandarin); *fucsia* (Spanish)

Scientific Family Name:
Onagraceae

Common Family Name:
willowherb family

Natural Distribution:
About 105 Fuchsia species grow naturally in Central and South America, Tahiti and New Zealand

USDA 6a–9b (AUS 1–2)

How to Grow: Best grown on neutral to mildly acidic or alkaline soil; keep evenly moist and mulch; fertilise every 4–6 weeks or include slow-release fertiliser or compost during planting; propagate from softwood or semi-hardwood cuttings, or from seed.

Drawbacks: Heat and dryness cause improper bud development and flowers to fall; insect pests include whiteflies, aphids, thrips, mealybugs, spider mites and scale; fungal diseases include rush, blight and others.

Plant Description:

- These grow as evergreen to deciduous shrubs (some bearing tubers) or small trees whose branches are trailing, spreading, loose, or upright and compact.
- The simple, stalked leaves are opposite or whorled, with 3–5 leaves per node. The edges of the leaf blade varies from being toothed to smooth. The leaf colour varies from golden-yellow to various shades of green. Young foliage is sometimes tinged red.
- Its pendulous, lantern-shaped, bird-pollinated flowers may grow solitarily, or in most cultivars, borne profusely in axillary or terminal branched or unbranched flowering shoots. There are 4 sepals and 4 petals, with the stamens, stigma and style protruding from the petals. Flowers may be single, semi-double or double, with great variation in shape, size, colour and shading.
- The fruit is an edible berry, containing numerous tiny seeds within the juicy flesh.

Pendulous, lantern-shaped bloom of *Fuchsia* cultivar

 Photos: (above and opposite) Hugh Tan Tiang Wah

Fuschia cultivar plant

Fuschia cultivar flower

Folklore, Beliefs, Uses and Other Interesting Facts

- According to feng shui beliefs, the profusely flowering fuchsias are considered highly auspicious because their flowers resemble red lanterns, and are prized for their *yáng* (male) energy. This is in spite of the taboo attached to plants with pendulous parts, such as the weeping willow (*Salix* species), that are seen to represent sadness. The Chinese hang lanterns around the home to welcome the Chinese New Year and herald good fortune. Hence, the lantern-shaped fuchsia bloom is regarded as a symbol of good tidings.
- Another feng shui belief is that the red- and magenta-flowered fuchsias represent the fire element, given their vibrant colours. They are, thus, deemed to be auspicious plants to grow in the front of homes, south, southwest and northeast sectors of one's garden. They are, however, inauspicious for the west and northwest sectors, and neutral for the rest.
- Fuchsias, especially hybrids such as *Fuchsia hybrida* (parents probably being *Fuchsia megallenica* and *Fuchsia fulgens*) and involving hybridisation with other species, are mainly cultivated as ornamental plants in temperate regions. They do not do well in the tropics as they grow best in areas with day temperatures of 15–21°C (59–69.8°F) and night temperatures of 10°C (50°F) or lower. A few species are used medicinally.

Gladiolus species & hybrids

gladiolus • sword lily

Scientific Species Name:
***Gladiolus* species** and **hybrids**
(Latin *gladiolus*, small sword, referring to the sword-like leaves of these plants)

Common Species Name:
corn-flag, gladiolus, sword flag, sword lily (English); *glaïeul* (French); *Gladiole* (German); *jiàn lán* [剑兰], *táng chāng pú* [唐菖蒲] (Mandarin); *estoque*, *gladiola*, *gladiolo*, *gladíolo* (Spanish)

Scientific Family Name:
Iridaceae

Common Family Name:
iris family

Natural Distribution:
Africa, Madagascar and Eurasia; there are about 260 *Gladiolus* species; cultivars of hybrids are widely grown; the common, large-flowered hybrids being the products of interbreeding between four or five species, followed by breeder selection, these being called *Gladiolus* x*gandavensis*, *Gladiolus* x*lemoinei*, and *Gladiolus* x*hortulanus*

USDA 7–10 (AUS 1–4)

How to Grow: Medium maintenance perennial plant; for best growth, dig the corm out annually and replant in the next growing season; humus-rich, medium moisture, well-drained soils except heavy clay; propagate by plant divisions.

Drawbacks: Diseases include *Botrytis* rot, crown rot, rust, wilt and mosaic virus; insect pests include aphids, mealybugs, spider mites and thrips.

Plant Description:

- Perennial, branched or unbranched herbs that grow from corms.
- The unstalked, alternate, 2-rowed, 1–9 leaves are sword-like or lance-shaped.
- The somewhat unstalked, somewhat fragrant, bilaterally symmetrical flowers grow in long, erect shoots. The 6 tepals are joined basally into a tube, and are in white, cream, yellow, orange, red, pink, green-lavender and purple, usually with contrasting coloured markings. The 3 outer tepals are narrower, with the dorsal tepal of the 3 inner ones the largest, arching over or hooding the stamens. The style overarches the stamens, and is divided into 3 thread-like branches.
- The fruit is a slightly bloated, softly cartilaginous capsule which usually contains many, broadly winged seeds, or, rarely, a few wingless seeds.

Gladiolus cultivar flowers

 Photos: (above and opposite) Hugh Tan Tiang Wah

Gladiolus cultivar flower sprays

Gladiolus cultivar flower

Folklore, Beliefs, Uses and Other Interesting Facts

- Gladioli (plural for gladiolus) are believed to be auspicious plants for the Chinese during their New Year. Red or yellow flowers present on each successive joint of the elongated flowering shoot symbolise growth in steady steps. The plant is therefore sometimes known as *bù bù gāo shēng* [步步高升] in Mandarin, which means 'consistent growth and improvement'.
- The Peranakans (descendants of Chinese immigrants in the Straits Settlements of the British, including parts of Malaysia [Penang, Province Wellesley, Malacca] and Singapore), and the Chinese of Singapore and Peninsular Malaysia, believe that by displaying flowering leafy shoots of gladioli in the home, the sword-shaped leaves help to ward off misfortune or evil and prevent spirits from harming members of the family.
- Based on feng shui beliefs, gladioli should be planted according to their flower colour. Red-, pink- or magenta-flowered plants are auspicious if planted in the south, southwest and northeast of the garden; inauspicious in the west and northwest; and neutral for the rest. For yellow- or orange-flowered plants, they are auspicious if planted in the southwest, west and northwest; inauspicious in the north; and neutral in the rest. For white-flowered plants, they are auspicious if planted in the west, northwest and north; inauspicious for the house front (white is the colour of death and mourning), east and southeast; and neutral for the rest. For purple-flowered plants, they are auspicious if planted in the north, east and southeast; inauspicious in the south; and neutral for the rest.
- Scholars suggest that one species of gladiolus, *Gladiolus italicus*, bears the expression of grief because its lower petals have markings resembling "*AI AI*", Greek for "alas, alas". As such, in Greek mythology, it is thought to be the plant which sprang from the blood of Hyakinthos, the youth accidentally killed by Apollo when they were playing discus together, and not the hyacinth (*Hyacinthus orientalis*).
- Gladioli are cultivated as ornamental plants or for cut flowers. There was a boom in gladioli in 1912, although not of the same intensity as that for tulips—the Tulipomania of 1636–1637—in the Netherlands.

*H*eavily Fruiting Plants

The more common plants cultivated for their auspiciousness are included below.

Scientific Name	Common Names	Fruit Colour
Capsicum annuum **cultivated varieties**	chilli (English); *piment* (French); *Chili Pfeffer* (German); *mirch* (Hindi); *cabai*, *cili* (Malay); *là jiāo* [辣椒] (Mandarin); *chile* (Spanish)	Greenish yellow, cream, orange, red, purple
Capsicum frutescens	bird's-eye chilli, chili pepper, goat pepper, hot chili, pungent pepper, spur pepper (English); *piment des oiseaux*, *piment enragé* (French); *Ziegenpfeffer*, *Vogelpfeffer*, *Roter Pfeffer*, *Chili Pfeffer* (German); *lalmirch*, *lalmirchi*, *lankamirchi*, *mirch* (Hindi); *cabai burung*, *cabai Melaka*, *cabai merah*, *cabai rawit*, *cili padi*, *lada api*, *lada merah* [Malaysia] (Malay); *là jiāo* [辣椒], *mǐ jiāo* [米椒], *xiǎo là jiāo* [小辣椒], *yě jiāo zǐ* [野椒子], *yě là zǐ* [野辣子] (Mandarin); *ají*, *chile*, *guindilla*, *pimienta picante* (Spanish)	Yellow, orange, red
×Citrofortunella microcarpa	See page 38	Orange
Citrus reticulata	See page 42	Orange
Fortunella japonica or ***Fortunella margarita***	See page 76	Orange
Lycopersicon esculentum **cultivated varieties**	garden tomato, love apple, tomato (English); *tomate* (French); *Tomate* (German); *tamatar* (Hindi); *terung masam*, *tomato* (Malay); *fān qié* [番茄] (Mandarin); *tomate* (Spanish)	Orange, red
Solanum pseudocapsicum	Jerusalem cherry, Madeira cherry, Madeira winter cherry, winter cherry, (English); *Jerusalemkirsche*, *Korallenstrauch* (German); *shān hú yīng* [珊瑚樱/珊瑚櫻] (Mandarin); *cereza de Jerusalén*, *cereza de Madeira* (Spanish)	Orange, red
Spondias dulcis **(=** ***Spondias cytherea*****)**	dwarf ambarella, golden apple, great hog plum, hog plum, Jew plum, makopa, Otaheite apple, Polynesian plum, Tahitian quince, wi tree, yellow plum (English); *casamangue*, *pomme de cythère*, *pommier de cythère*, *prune de cythère*, *prunier de cythère*, *prunier d'Amérique* (French); *Goldpflaume*, *Süsse Mombinpflaume*, *Suesse Mombinpflaume*, *Tahitiapfel*, *Tahiti-Apfel* (German); *amra* (Hindi); *kedongdong* (Malay); *bīng láng qīng* [槟榔青], *rén miàn zǐ* [人面子] (Mandarin); *ambarella*, *jobo de la India* (Spanish)	Yellow

 Photos: (opposite) Hugh Tan Tiang Wah

Capsicum frutescens fruiting plant

USDA >11 (AUS >5)

How to Grow: Please see under the species' own treatments in this book, or check with other references, if not. In general, these plants, particularly annuals, are purely for ornamental purposes, so are maintained only for a few weeks. In general, keep in partial shade and water regularly to prevent dehydration.

Drawbacks: Plants, especially annuals, will usually die after a few weeks; purchase new plants to replace them when they die or become unhealthy.

Plant Description:

- Potted plants or small trees bearing numerous fruits, usually orange or red in colour.

Solanum pseudocapsicum fruits on leafy branches

Folklore, Beliefs, Uses and Other Interesting Facts

- The Chinese believe that a plant heavily laden with fruit is auspicious, especially vigorously growing plants with bright orange fruit, like gold, representing great good fortune, happiness, wealth and general prosperity, and particularly when the fruits develop during the Chinese New Year, they will attract prosperity and good luck. Red fruits are also auspicious and indicative of life and vigour, as red is the colour of blood.
- Based on feng shui beliefs, such plants should be placed in view of the front door or in the east of the garden (the Big Wood zone) to be auspicious. If planted in the ground, such plants, particularly trees, have room to grow and symbolise unlimited growth. Potted plants will have more limited growth, so they need to be nurtured by adding in compost and nutrients, and pruned to stimulate more growth. Immediately replace plants that are unhealthy or sickly looking or dead, which are highly inauspicious, with healthy, vigorous plants.

Lycopersicon esculentum cultivar fruiting plant in wrapper for sale

Hedera helix ivy • common ivy

Scientific Species Name:
Hedera helix
(Latin *hedera*, this species; Greek *helix*, anything spiral-shaped, referring to the climbing habit of this plant)

Common Species Name:
common ivy, English ivy, ivy (English); *lierre* (French); *Efeu* (German); *vrikshalata* (Hindi); ivy (Malay); *cháng chūn téng* [常春藤] (Mandarin); *hiedra* (Spanish)

Scientific Family Name:
Araliaceae

Common Family Name:
ivy family

Natural Distribution:
Europe, Mediterranean and West Asia

USDA 5–9 (AUS 1–3)

How to Grow: Fast growing, temperate climber; best in rich, moist, well-drained, acidic to slightly alkaline soil; low salt tolerance; medium drought tolerance; propagate by stem cuttings (stem roots with ground contact), air layering or grafting.

Drawbacks: To avoid fungal diseases (e.g., *Rhizoctonia*), well-circulated air and well-drained soil are the best conditions, so do not overwater; contains toxic saponins poisonous to mammals and molluscs; handling it may cause skin irritation or allergies from the sap; ivy roots can damage the walls on which they climb; can aggressively spread and become an invasive weed; scale insects are a major pest.

Plant Description:

- Long-lived, perennial, evergreen, woody climber, which climbs by roots which attach to the support structure.
- Leaves are alternate and simple, dark green or variegated ivory (some cultivated varieties) and of two kinds. Those of non-flowering branches are 3–5-lobed and up to 10 cm (3.9 in) long. Those of flowering branches are egg or diamond-shaped, with unlobed margins.
- The flowers grow in clusters at the branch tips.
- The round fruit is up to 8 mm (0.3 in) across, deep bluish purple to black when ripe, containing 2–5 seeds.

 Photos: (opposite) Gertrude Kanu

Hedera helix climbing plant showing leaves of sterile branches

Flower clusters at branch tips

Ripening fruits

Folklore, Beliefs, Uses and Other Interesting Facts

- Ivy is an evergreen associated with Christmas in the northern hemisphere, possibly for being symbolic of undying life during winter when most plant life is dormant. It is used in magic rites to ensure the return of plant growth, so sacred buildings in Europe and Western Asia were decorated with Christmas evergreens for rituals of the winter solstice. Another possibility is that, for practical reasons, evergreens were the only plants available during the northern hemisphere winter and were used to add colour to homes. In the British Isles, Christmas greenery should not be brought indoors before Christmas Eve and should be removed before the New Year or the 12th day of Christmas (6th of January). During Christmas, ivy is considered lucky for women and holly for men. Ivy is usually forbidden indoors at any time other than Christmas as it is considered unlucky then.
- In the British Isles, ivy growing on a house protects the occupants from witchcraft and evil. However, there will be misfortune if it dries up or falls off suddenly. In Wales, the latter is believed to predict the transfer of house ownership as a result of financial losses or some other reason.
- In the language of flowers, ivy symbolises fidelity, friendship, fidelity in friendship, matrimony or reciprocal tenderness.
- The many cultivated varieties of ivy are grown as ornamental climbers for walls, as pot plants, as standards (grafted on ×*Fatshedera* rootstocks) or as trailers for ground cover.
- Ivy is used in traditional medical treatments for numerous ailments, including corns (vinegar-soaked leaves attached to the corn) and as an antidote to the effects of alcohol (drinking vinegar into which ivy berries have been dissolved before drinking alcohol then drinking water which has been boiled with the bruised leaves afterwards). The ivy is often depicted in the wreath of Bacchus (the god of wine of ancient Greece and Rome) or as a sign for taverns.
- Ivy is naturalised throughout the temperate world and one of the world's most invasive weeds where it was introduced, such as in Australia, Hawaii and New Zealand.

Hordeum vulgare barley • malting barley

Scientific Species Name:
Hordeum vulgare
(Latin *hordeum*, barley;
Latin *vulgare*, common)

Common Species Name:
barley, common barley, cultivated barley, hooded barley, malting barley (English); *orge, orge de printemps, orge d'hiver, orge vulgaire* (French); *Gerste, Kulturgerste, Saatgerste* (German); *jau* (Hindi); *barli, beras Belanda, padi Belanda* (Malay); *dà mài* [大麦/大麥] (Mandarin); *cebada, cebada común, cebada cultivada* (Spanish)

Scientific Family Name:
Poaceae or Gramineae

Common Family Name:
grass family

Natural Distribution:
Probably native to the Middle East, from Afghanistan to northern India

USDA 1–11 (AUS 1–6)

How to Grow: An annual grass that can be grown from sub-Arctic to high mountains of the tropics; preferring climates with cool and moderately dry seasons; not well suited for hot, humid climates; rainfall from 200–1,000 mm (7.9–39.4 in) per year; very drought-tolerant, but intolerant to waterlogging; soil pH 4.5–8.3, preferring deep loams with pH 7–8; somewhat salt-tolerant; propagation by seed (usually a spring crop in temperate climates).

Drawbacks: When grown in a field as a monoculture, it is prone to weeds and several diseases caused by viruses (such as barley yellow dwarf virus and barley stripe mosaic virus) and fungi (such as powdery mildew, spot blotch, scald, scab, rusts). Aphids are the major insect pests.

Plant Description:

- Erect, annual grass up to 60–120 cm (23.6–47.2 in) tall that tillers freely.
- Leaves are few, narrowly lanced-shaped, up to 25 cm (9.8 in) long and 1.5 cm (0.6 in) broad, arising alternately from the erect stem.
- Flowers are arranged in a dense spike-like compound inflorescence composed of spikelets, each of which are sessile and arranged in groups of threes at each joint of the inflorescence axis in 2 rows.
- The single-seeded fruit is commonly known as the grain of barley.

Hordeum vulgare grains

Photos: (above) Steve Hurst @ USDA-NRCS PLANTS Database; (opposite) Christian Guthier

Hordeum vulgare field in Wheatley, Oxford, UK

Compound infrucescence

Folklore, Beliefs, Uses and Other Interesting Facts

- Barley is one of the more prominent plants in the Bible. In their seminal work on biblical plants, Moldenke and Moldenke recorded that barley was mentioned in the bible 32 times either as a plant growing in the field or as products made from it.
- In Deuteronomy 8:7–8 (King James Version), the people of Israel were promised: "....a good land....A land of wheat, and barley, and vines, and fig trees, and pomegranates; a land of olive oil, and honey", so these have a special place in Judaism as the seven plants of the Bible.
- In Biblical times, barley was food for the poor. Barley grains were ground and made into round cakes. It was also grown as fodder for horses.
- Today, barley is used mainly for livestock feeding and in the brewing of alcoholic beverages.

Hyacinthus orientalis hyacinth

Scientific Species Name:
Hyacinthus orientalis
(*Hyacinthus*, after the Greek prince of the same name, and Latin *orientalis*, originating in the Orient or Eastern)

Common Species Name:
common hyacinth, Dutch hyacinth, garden hyacinth, hyacinth, Roman hyacinth (English); *jacinthe* (French); *Hyazinthe* (German); *sumbul* (Hindi); *fēng xìn zǐ* [风信子] (Mandarin); *jacinto* (Spanish)

Scientific Family Name:
Hyacintheaceae

Common Family Name:
hyacinth family

Natural Distribution:
Northeastern Mediterranean but naturalised in Europe

USDA 7–9 (AUS 1–3)

How to Grow: A temperate, bulbous species which prefers cool climates; with numerous cultivated varieties; do not overwater; soil pH of 6.6–7.5 (slightly acid to slightly alkaline); plant 15–22 cm (5.9—8.7 in) apart in autumn with the base of the bulbs 15–18cm (5.9—7.1 in) from the soil surface, and deeper in the cooler extremes of its growing range; propagation by division of the bulbs, cutting crosswise slits or gouging the bulb base to stimulate new bulb growth or through tissue culture; seed germination when possible, usually does not produce plants true to type.

Drawbacks: Not suitable for the tropics where air-flown imports tend to die after a while, even when grown in cool shade; all plant parts are poisonous and handling the plant with bare hands may result in an allergic rash or skin irritation; bulbs may be prone to rot.

Plant Description:

- A perennial, bulbous flowering herb growing to about 30 cm (11.8 in) tall.
- Bears 4–6, strap-like, lance-shaped leaves, each of which grows up to 15–25 cm (5.9—9.8 in) long from the bulb.
- Up to 40 fragrant, red, orange, pink, yellow, pale blue to violet, lilac or white flowers that are arranged in a flowering shoot. Roman hyacinth (*Hyacinthus orientalis* var. *albulus*) has less crowded blue or white flowers than the typical variety (*Hyacinthus orientalis* var. *orientalis*), which has more flower colours. Numerous cultivated varieties are available which are diploid, triploid or with higher ploidy levels. 'Double' flowered cultivars are also available.

Photos: (opposite top) Judy Malley; (opposite bottom) Lauren Bansemer

Inflorescence of *Hyacinthus orientalis* cultivar

Hyacinthus orientalis cultivar plant and inflorescence

Folklore, Beliefs, Uses and Other Interesting Facts

- *Hyacinthus* is Greek for the hyacinth. It is a word derived from the Thraco-Pelasgian language and originally the name of a god associated with the regrowth of plants. In Greek mythology, Hyakinthos is a beautiful youth who was killed accidentally while trying to catch a discus thrown by Apollo. A plant was said to have sprung from Hyakinthos's spilled blood, and henceforth named 'hyacinth'. (Scholars suggest instead that the actual species is probably *Gladiolus italicus*, whose lower petals have markings resembling "*AI AI*", Greek for "alas, alas".)
- In temperate regions, the common hyacinth flowers in spring. Hence, to the Chinese, the flowering of the hyacinth symbolises the coming of spring and the arrival of the Chinese New Year. Other blooming temperate plants used to symbolise the coming of spring include the narcissus and the cherry/plum. Red, pink or violet flowers are preferred because of their association with life, fertility and good luck. White flowers are to be avoided because it is the colour of mourning and death.
- Based on feng shui beliefs, red- or pink-flowered hyacinths are auspicious if planted at the south, southwest or northeast zones of the garden, or at the front of the house, and inauspicious in the west or northwest zones, but neutral for the rest. Violet-flowered hyacinths are auspicious in the north, east or southeast zones of the garden; inauspicious in the south; and neutral for the rest. For yellow- or orange-flowered plants, they are auspicious if planted in the southwest, west and northwest; inauspicious in the north; and neutral in the rest. For white-flowered plants, they are auspicious if planted in the west, northwest and north; inauspicious for the house front (white is the colour of death and mourning), east and southeast; and neutral for the rest.
- Bulbous plants such as hyacinth are considered good feng shui plants because a flowering bulb is considered buried gold, staying hidden until the appropriate time to burst forth. This is particularly auspicious if blooming occurs during the Chinese New Year. To have hyacinths in the home during the winter months, when *yīn* energy predominates, is good as they bring in lively *yáng* energy for better harmony and balance.
- In the language of flowers, the hyacinth represents play or games (sport), benevolence, jealousy, chagrin, or means "You love me and give me death". The common hyacinth symbolises a game because Hyakinthos died whilst playing a game.
- In the popular culture of the Czechs, potted hyacinths are used to decorate the house during Christmas.
- The flowers are used for scent-making in Southern France (phenyl acetaldehyde and other sensorially important substances).

Hypericum species St. John's wort

Scientific Species Name:
***Hypericum* species**
(Greek *hypereikon*, St. John's wort; derived either from the Greek *ereike*, heath, or Greek *hyper*, above, and Greek *eikona*, icon or picture; see opposite for the detailed explanation of this name)

Common Species Name:
hypericum, St. John's wort (English); *millepertuis* (French); *Johanniskräuter*, *Hartheu* (German); *jīn sī táo* [金丝桃/金絲桃] (Mandarin); *hipérico*, *hierba de San Juan*, *yerba de San Juan* (Spanish)

Scientific Family Name:
Hypericaceae (formerly placed in the Clusiaceae or Guttiferae)

Common Family Name:
St. John's wort family
(formerly in the pitch-apple or mangosteen family)

Natural Distribution:
Almost cosmopolitan but do not occur in arctic, desert or tropical lowland areas. There are about 460 species worldwide.

USDA 3–8 (AUS 1–2)

How to Grow: This description applies to perforate St. John's wort or Klamath weed (*Hypericum perforatum*); for other species, please check from other references; regular watering; dry or well-drained soil; propagate by seed, and plants frequently self-sow, also by plant division.

Drawbacks: Klamath weed has become an invasive weed species in some countries where it has been introduced (it is native to the Atlantic Islands, Europe, Northwest Africa, Southwest to Central Asia, including China); the plant is poisonous to livestock through increasing photosensitisation, especially in sunny localities such as Australia, Southwest USA or Iraq.

Plant Description:

- St. John's worts or hypericums refer to all *Hypericum* species.
- They range from small annual or perennial herbaceous plants of under 1 m (3.3 ft) tall, to evergreen of deciduous shrubs and small trees.
- The oval-shaped leaves are opposite or whorled, and the leaf blades are gland-dotted (contain scattered glands which make them appear perforated since the glands are translucent when held up to a light source).
- The 5 (rarely 4)-sepalled and 5-petalled flowers are star or cup-like. The petals vary from golden to lemon yellow and are rarely white. There are numerous stamens in 5 (rarely 4) bundles. The pistil consists of an ovary and 2–5 styles, tipped by a stigma each.
- The fruit is usually a dry 3–5-celled capsule which usually splits to disperse its numerous, tiny seeds. However, in some species the fruit is fleshy, almost berry-like.

 Photos: (opposite top) Alfred Wein; (opposite middle) Peggy Greb, ARS, USDA; (opposite bottom) Eric Hunt

Hypericum perforatum plants growing wild

Hypericum perforatum plantlets

Hypericum perforatum flower

Folklore, Beliefs, Uses and Other Interesting Facts

- *Hypericum* can be interpreted to be derived from two Greek words: *hyper*, above, and Greek *eikona*, icon or picture. This probably refers to the practice of hanging the flowers of this plant above pictures or icons to ward off evil in the house during St. John's Day. In revenge, it is believed the Devil pierced the leaves of this plant with a needle (the gland-dotted appearance of the leaf blade giving this impression).
- The hypericum plant was used in pre-Christian rituals in England to ward off evil and to turn away witches. It was once called *Fuga Daemonum*, meaning (literally) "escape evil spirit" in Latin, to reflect that it drove away evil spirits and prevented ghosts from entering or haunting houses. Yellow flowers are generally associated with magical properties but their powers were thought to be greatest in midsummer, coincidentally when St. John's wort blooms.
- Its common name, St. John's wort, comes from the annual flowering and harvesting of the plant on St. John's Day (24 June), which celebrates the birth of St. John the Baptist. The saint and the plant are further linked by the folklore that the red spots of St. John's wort are his spilled blood, appearing annually on 27 August, the anniversary of his beheading.
- In the British Isles, Roy Vickery cites numerous reports that St. John's wort was highly regarded for protection against demons, fairies and the devil, warding off second sight, the 'Evil Eye', death and lightning, and to ensure peace, plenty and prosperity.
- According to the language of flowers, St. John's wort signifies superstition, superstitious sanctity, or originality.
- Based on feng shui beliefs, yellow-flowered plants, such as most St. John's worts, are auspicious if planted in the southwest, west and northwest; inauspicious in the north; and neutral in the rest.
- *Hypericum perforatum* is commonly used in the present-day as a herbal cure for depression. Scientific studies have shown that herbal preparations of this species show greater efficacy than placebos and comparable efficacy to prescription antidepressants and with fewer side effects. A more recent 2006 study confirmed that the effect of this plant is clinically significant but only on patients with minor depression and not patients with dysthymia and major depression. A National Institute of Health-funded study further confirmed that the species is ineffective in treating moderate to severe cases of major depression.

Ilex species & hybrids **holly**

Scientific Species Name:
Ilex **species** and **hybrids**
(Latin *ilex*, ancient name of the evergreen holm oak [*Quercus ilex*], referring to the similarity of holly leaves to those of this plant; there are about 400 *Ilex* species, but *Ilex aquifolium* and *Ilex* x*altaclerensis* are most commonly cultivated)

Common Species Name:
holly (English); *houx* (French); *Stechpalme* (German); *holi* (Malay); *dōng qīng* [冬青] (Mandarin); *acebo* (Spanish)

Scientific Family Name:
Aquifoliaceae

Common Family Name:
holly family

Natural Distribution:
Cosmopolitan

USDA 6–8 (AUS 1–2)

How to Grow: The growing conditions and drawbacks listed here pertain mostly to those of *Ilex aquifolium*, so for the numerous other species or hybrids, please check with other references (e.g., the website of the Holly Society of America (http://www.hollysocam.org/); hardy plant; well-drained, light soil of pH 4.5–6; do not overwater (intolerant of flooding) but water regularly; intolerant of high temperatures or humidity; propagation from semi-woody cuttings or seed.

Drawbacks: Plant parts are poisonous; spiny leaves require care during pruning and other maintenance work; fungal diseases include leaf spot, blight, root rot, *Sphaeropsis* knot and tar spot; nematodes may also attack roots; insect pests include leaf miners, Southern red mite and scale.

Plant Description:

- This description applies to *Ilex aquifolium*.
- An evergreen shrub or tree 2–15 m (6.5–49.2 ft) tall, with heavily branched, conical crown.
- The alternate, stalked leaves have leathery leaf blades that are 5–12 cm (2.0–4.7 in) long and 2.5–5 cm (1–2.0 in) wide, dark, glossy green, wavy, with usually spiny margins which are thickened. There are also variegated leafed cultivars.
- Its fragrant flowers occur in clusters in the previous year's growth and petals are white to pinkish-white and joined basally. Holly plants are either male or female. Male flowers have 4 stamens and no distinct pistil, whereas female flowers have a pistil and 4 staminodes, which produce no pollen.
- The fruits are bright red, round drupes 8–10 mm (0.3–0.4 in) across.
- There are 4 seeds per fruit, each enclosed in a tough covering formed from the inner fruit wall.

 Photos: (opposite top) Scott Bauer, ARS, USDA; (opposite middle) Anna Buxton; (opposite bottom) Peter Birch

Ilex 'Scepter', a cultivar introduced by the US National Arboretum in 1999

Ilex aquilifolium fruits

Ilex aquilifolium male flower and buds in a cluster

Folklore, Beliefs, Uses and Other Interesting Facts

- The holly was held sacred by the druids of ancient Britain and Gaul, and are widely planted to protect against witchcraft, mad dogs and evil beings. Holly has also been traditionally associated with protection from lightning and so is planted near the house. It was considered taboo to cut down a holly tree, but it is permissible to cut branches for winter fodder or decoration.
- Before the coming of Christianity, European pagans harvested wild hollies and placed them in their homes, so that the tiny imaginary people of the woodland could seek refuge from the harsh winters. It is also possible that the use of the holly to decorate homes in winter is associated with the Roman Saturnalia (a feast for the god Saturn on 17 December). However, when Christianity was widely adopted by most of Europe, the use of holly was banned by the Christian council because of its pagan origin. Despite the ban, holly continued to be used as a decoration during festive seasons. The reason for its use slowly changed from that based on mythical pagan beliefs to that based on Christian symbolism. It was widely believed that the spiny leaves and red berries represented the crown of thorns and the blood of Christ respectively. The Germans named the holly, amongst other plants, *Christdorn*, Christ's crown of thorns, believing that the fruits of the holly were originally white but became red when stained with Christ's blood.
- Holly is an evergreen associated with Christmas in the northern hemisphere, possibly for its symbolism of undying life during winter when most plant life is dormant. It is used in magical rites to ensure the return of plant growth, so sacred buildings in Europe and Western Asia were decorated with Christmas evergreens for rituals of the winter solstice. Another possibility is that, for practical reasons, evergreens were the only plants available in the northern hemisphere winter and were used to add colour to homes. Christmas greenery should not be brought indoors before Christmas Eve and should be removed before the New Year or the Twelth Day of Christmas (6 January). Holly, especially a heavily fruited one, is the plant most associated with Christmas throughout the British Isles. In some parts of England, it is even called 'Christmas'. However, it is considered unlucky to bring this plant indoors except during the Christmas season.
- In the British Isles, holly is considered a lucky plant because it is evergreen with red fruits, symbolic of unceasing life. Male (or prickly) holly is good luck for men and female (or the smooth-margined, variegated leafed type) holly is good luck for women.
- In modern Europe and North America, the holly is seen primarily as a symbol of Christmas. The glossy, dark green leaves and bright red fruits complement each other perfectly—a beautiful and meaningful reminder of the joy that Christmas brings.
- Spiny plants, such as holly, are considered 'bad' feng shui plants because their sharp spines symbolise numerous 'poison arrows'—hostile, inauspicious energy which harm the residents. For this reason, holly must not be placed inside the home. If one is a great fan of the holly, the compromise is to grow it as a friendly sentinel by placing it in the garden near the gate or outside the wall or fence, to guard the entrance to the home.

Jasminum sambac Arabian jasmine

Scientific Species Name:
Jasminum sambac
(*Jasminum* is the late medieval Latin version of the Persian *yasmin* or *yasmimin*, the name of this species; *sambac*, after the Persian plant name, *zambac*, the name of this species)

Common Species Name:
Arabian jasmine, jasmine, zambac (English); *jasmin, jasmin d'Arabie* (French); *Jasmin* (German); *bela, chameli, magadhi* (Hindi); *melati, melur* (older spelling *melor*) (Malay); *mò lì huā* [茉莉花] (Mandarin); *jazmin, jazmin de Arabia* (Spanish)

Scientific Family Name:
Oleaceae

Common Family Name:
olive family

Natural Distribution:
Probably India; now widely cultivated in temperate subtropical and tropical countries

USDA ≥9 (AUS ≥3)

How to Grow: Cold-tolerant, subtropical species; can tolerate being left in the ground through winter; moist, but not waterlogged soil, especially during hot spells; nitrogen fertiliser application increases flower production; propagate by rooting semi-woody twigs or leaf cuttings (one mature leaf and an axillary bud), or layering.

Drawbacks: Prone to insect attack, especially in the tropics, so must spray heavily; must be pruned regularly or becomes very straggly and untidy.

Plant Description:

- Small, evergreen erect or climbing shrub.
- The opposite, stalked leaves, have leaf blades that are 2.5–9 cm (1–3.5 in) by 2–6.5 cm (0.8–2.6 in), shiny green, oval or egg-shaped, with smooth, wavy margins.
- Strongly fragrant flowers, 1.5–2.5 cm (0.6–1.0 in) across, are arranged in clusters of 3 to many flowers. The petals are joined together in a tube at the base with 5 ('single' flowers) to many free lobes ('double' flowers), which turn from white to slightly off-white or pink with age.
- The fruit is a black berry.

Jasminum sambac
double flower

 Photos: (above) Giam Xingli; (opposite top and bottom) Hugh Tan Tiang Wah

Jasminum sambac flowering and leafy branch

Flower offerings with Saraswathi (Goddess of Wisdom) idol

Folklore, Beliefs, Uses and Other Interesting Facts

- Held sacred to the Hindu god, Vishnu the Preserver, the Arabian jasmine flowers are used as votive offerings. The double flowers are often strung up in garlands and placed at altars for worship. Garlands are also often used by the Hindu community to welcome honorable guests from both within and without the community during special occasions.
- Southeast Asian Malays believe that its fragrance attracts spirits, so it is used in seances to call upon spirits for their help or advice. On the other hand, the flowers are also used as part of rituals to exorcise spirits which cause diseases.
- The Arabian jasmine flower is also a favourite among the Malays because of its strong fragrance, and women wear the flowers in their hair. Water soaked with Arabian jasmine flowers is used for washing faces and coconut oil soaked with these flowers is used for hair grooming. Arabian jasmine scent is also thought to relieve headaches.
- Malaysian Indians believe that snakes are attracted to Arabian jasmine fragrance, so they prune off the lower branches to reduce cover.
- Based on feng shui beliefs, white-flowered plants like the Arabian jasmine are auspicious if planted in the west, northwest and north, inauspicious for the house front (white is the colour of death and mourning), east and southeast, and neutral for the rest.
- In the language of flowers, the white jasmine represents amiability, candour, or grace and eloquence.
- The Arabian jasmine is the national flower of Indonesia (*melati* in Bahasa Malaysia [Malay]) and the Philippines (known as the *sambaguita* in Tagalog).
- Dried flowers are added to tea (*Camellia sinensis*) to produce the fragrant jasmine tea drunk by the Chinese and Javanese.
- Jasmine oil is extracted for use in perfumes.
- Arabian jasmine leaves, flowers or roots are used medicinally as folk cures for fever, skin complaints, sprains, fractures, bronchitis and asthma, and as a decongestant, sedative or anaesthetic. Women in India, Indonesia, Malaysia and the Philippines apply a poultice of the bruised leaves and flowers onto their breasts to stop milk secretion after childbirth.

*J*uniperus species & hybrids juniper

Scientific Species Name:
***Juniperus* species** and **hybrids**
(Latin *Juniperus*, juniper;
there are about 50–70 species)

Common Species Name:
juniper (English); *genièvre* (French);
Wacholder (German);
hauber (Hindi);
cì bái [刺柏] (Mandarin);
enebro (Spanish)

Scientific Family Name:
Cupressaceae

Common Family Name:
cypress family

Natural Distribution:
Widely distributed throughout the northern hemisphere, from the Arctic to Central America, East Africa, Himalayas and Taiwan

USDA 3–9 (AUS 1–3)

How to Grow: A generalised account is provided here, so for specific species' needs, please check other references; hardy, low-maintenance plants; most types of well-drained soils; generally flooding intolerant; may be drought tolerant when established; propagate by seed (may take years to germinate!), stem cuttings, layering or grafting.

Drawbacks: Generally hardy, with no serious pest or disease problems; may be susceptible to tip and needle blights, juniper blight, cedar-apple rust and related rust diseases, phomopsis twig blight, root rot may occur, particularly in wet, poorly drained soils, canker on bark or main stems; occasional insect pests include aphids, bagworms, scale, twig borers, webworms; acarid pests include spider mites; may have occasional dieback of foliage.

Plant Description:

- Evergreen, coniferous, low spreading shrubs to tall trees growing up to 40 m (131.2 ft) tall.
- The thin trunk bark exfoliates in long strips or scales.
- Leaves are in pairs and right angles to adjacent ones or in whorls of 3. In all juniper species, juvenile leaves are needle-like; adult leaves are needle-like or flattened on the stem and scale-like.
- Male (pollen producing) cones are yellow, egg-shaped or oblong, with 6–16 cone scales. Female (seed producing) cones are berry-like (not a true fruit since angiosperms, not conifers, produce fruit), round- or egg-shaped, with the cone scales joined and succulent. Birds usually disperse the seeds by eating the cone. Plants may produce both cone types (bisexual plants), or only male or female ones (unisexual plants).
- Seeds are 1–10 per female cone.

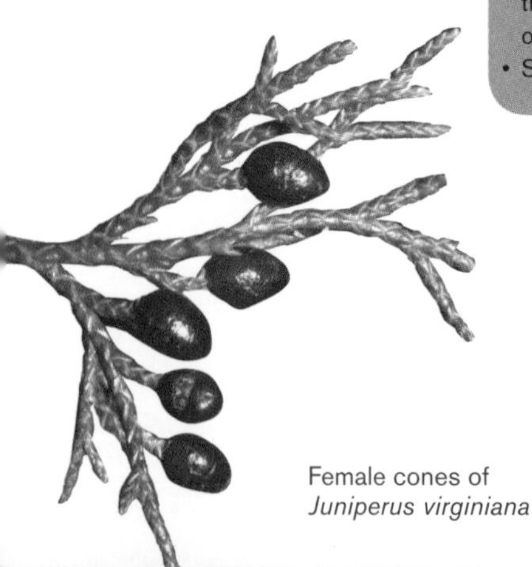

Female cones of
Juniperus virginiana

Photos: (above) Steve Baskauf; (opposite top) Peggy Greb, ARS, USDA; (opposite middle) Neil Gilham; (opposite bottom) Hugh Tan Tiang Wah

Juniperus californica bonsai plant

Juniperus species tree at Rock Canyon, west of Las Vegas, USA

Juniperus chinensis cultivar in a garden in Singapore

Folklore, Beliefs, Uses and Other Interesting Facts

- In ancient Canaan (a region which covered the present-day Israel, West Bank and Gaza, and the adjacent coastal lands and parts of Lebanon and Syria), the juniper was a symbol of the Canaanites' fertility goddess Ashera.
- In many countries, the juniper is considered a merciful and protective tree. An Italian legend has it that a juniper tree saved the life of the Infant Jesus by hiding him in its branches when Herod's pursuing soldiers came close. They only saw an old man (Joseph) walking with his wife (Mary), and so they walked on without stopping.
- In the Bible's Old Testament, as Elijah rested under a juniper tree after escaping from Queen Jezebel's vengeance, an angel visited and fed him (1 Kings 19:4–7, King James Version).
- During the Middle Ages, juniper branches were burned to dispel evil elements such as demons, devils, vermin and 'satanic' snakes from human settlements.
- Ancient Germans considered juniper sacred as it was the 'tree of life'. Juniper twigs were attached to their houses to protect them from demonic powers. Twigs were also given to the dead to help them on their journey to Valhalla.
- In the Western Isles of the British Isles, juniper branches placed in barns protect cows from the 'Evil Eye,' and juniper twigs were used to bind their legs during milking for the same reason.
- The juniper can be considered unlucky or lucky. In Wales, old juniper trees were allowed to die on their own because it was believed that person who felled a tree would die within a year or lose a family member. This taboo may have been started to protect this 'protective' tree!
- To dream of the juniper tree is unlucky, and if the dreamer is unwell, he is unlikely to recover. However, to dream of gathering juniper berries in winter foretells prosperity, and to dream of the berries themselves indicates future honours and success, or the birth of a son if one is married.
- The juniper is also considered a sacred and magical plant. Old German and Finnish tribes used juniper berries as an additive to their ritual beers. Beer was originally a ritual drink used as an offering to the gods or as a magical agent for altering everyday consciousness. Germanic tribes venerated beer, and during Germanic carousels, beer and mead (a drink made from honey) were consumed so that the gods would leave the heavens, come to earth and sojourn amongst the humans.
- Decoctions of juniper berries were said to give the gift of prophecy, while the cone was claimed to effect love magic and was used in necromantic exorcisms.
- Juniper is essential to the ritual and magical practices of central and northern Asia. Shamans of Siberian groups inhale juniper smoke to induce trance. Juniper smoke is also inhaled by people of the Himalayas prior to carrying out rituals, oracles, exorcisms, or healing.
- In the language of flowers, the juniper symbolises protection, asylum or succour, ingratitude or says "I live for thee."

Lagenaria siceraria bottle gourd

Scientific Species Name:
Lagenaria siceraria
(Greek *lagenos*, a flask, referring to the shape of the fruit of *Lagenaria* species; Greek *sikera*, an intoxicating spirit or liquor)

Common Species Name:
bottle gourd, calabash, calabash gourd, white-flowered gourd (English), *calebassier*, *calebassier grimpant*, *gourde bouteille*, *cougourde*, *courge siphon*, *courge massue* (French); *Flaschenkürbis*, *Flaschen-Kürbis*, *gewöhnlicher Flaschenkürbis*, *Kalebassenkürbis*, *Trompetenkürbis* (German); *dudhi* [*dudi*, *dodi*], *lokhi* [*lauki*] (Hindi); *labu botol* (Malay); *hú lú* [葫芦/葫蘆], *hú lú guā* [葫芦瓜/葫蘆瓜] (Mandarin); *acocote*, *cajombre*, *calabaza*, *calabaza vinatera*, *guiro amargo* (Spanish)

Scientific Family Name:
Cucurbitaceae

Common Family Name:
cucumber family

Natural Distribution:
Possibly originated in Tropical Africa, but now pan-tropical owing to widespread cultivation

USDA ≥11 (AUS ≥5)

How to Grow: Tropical climber; intolerant of frost; well-drained, light soils with fully decomposed manure or compost incorporated, pH 6–7; propagated by seed.

Drawbacks: Prone to disease if excessive soil moisture present; anthracnose, powdery mildew and basal rot (fungal diseases); fruit flies and leaf folders (insect pests); annual plant so needs to be regrown from seed each year.

Plant Description:

- Annual, non-woody climber of up to 10 m (32.8 ft) long or more. Hairs are present on all vegetative parts.
- The stalked, alternate leaves have oval-kidney-shaped or somewhat circular leaf blades with toothed margins. Tendrils, which are usually forked, grow from the leaf axils.
- The white-petalled flowers are produced singly at the leaf axils. Each plant bears male and female flowers. Male flowers have longer stalks and larger petals compared to the female's. There are 5 petals.
- The light green fruit is variably shaped according to cultivar–round, bottle or club-shaped–up to ≥1 m (3.3 ft) long.
- The numerous whitish or brownish seeds found in the ripe (dried) fruit are corky, broad and flat, or narrow and forked.

Bottle gourd adorned with Chinese New Year finery

Photos: (above) Hugh Tan Tiang Wah; (opposite) Giam Xingli

Folklore, Beliefs, Uses and Other Interesting Facts

- The bottle gourd fruit is featured in many Chinese folktales and pictorial depictions of deities and goddesses. A popular deity, the God of Longevity (*shòu xīng*, 寿星/壽星), is always depicted in paintings carrying a bottle gourd. It is believed that the bottle gourd contains the elixir of life thought to give immortality to the person who ingests it. Therefore, the bottle gourd is believed to bless a family with good health throughout the year. Fruiting bottle gourd potted plants or bottle gourd fruits are sold during the Chinese New Year as auspicious decorations.

Bottle gourd potted plants sold during the Chinese Lunar New Year

- In the Taoist pantheon of the Eight Immortals, *tiě guǎi lǐ* (铁拐李/鐵拐李, Iron-crutch Li) uses the bottle gourd to carry magic medicine. He also used it to revive the dead mother of one of his disciples, thus making this gourd a symbol of Taoist magic. It is placed in the bedroom to protect a healthy person against diseases and to improve the medical condition of one who is sick.
- The Chinese also believe that the fruit acts as a charm to protect against ghosts and evil spirits that may want to reside in their home. The potted fruiting vine is sold during the Chinese New Year as an auspicious plant for decoration.

Bottle gourd fruit

- In Japan, the fruit is sold in marinated strips and used in sushi. In Central America, the fruit is an important ingredient in a beverage, *horchata*. In Southern Africa, the young fruit is eaten as a vegetable. Its young shoots and seeds are also edible. The young green fruit is made into a syrup and applied on the chest to relieve pain. The dried fruit is used as cups, dippers, flasks, floats, hats, musical instruments, and in New Guinea as penis-sheaths.

Laurus nobilis bay tree • bay laurel

Scientific Species Name:
Laurus nobilis
(Latin *laurus*, name of this species; Latin *nobilis*, excellent, famous, notable or renowned)

Common Species Name:
bay laurel, bay leaf laurel, bay tree, Grecian laurel, laurel, sweet bay (English); *laurier* (French); *lorbeer* (German); *hab-el-gar* (Hindi); *laurel* (Malay); *xiāng yè* [香叶/香葉], *yuè guì* [月桂] (Mandarin); *laural* (Spanish)

Scientific Family Name:
Lauraceae

Common Family Name:
laurel family

Natural Distribution:
Eastern Mediterranean and Asia Minor

USDA 8–10 (AUS 2–4)

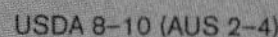

How to Grow: A mediterranean tree that is tolerant of light frost; grows best with temperatures from 8–27°C (46.4–80.6°F) and precipitation from 300–3,200 mm (11.8–126 in); well-drained, deep and fertile soils favour its growth, and waterlogging is not tolerated; soil pH range of 4.5–8.3 (acidic to alkaline); propagation by seed, air-layering, or stem or root cuttings.

Drawbacks: Can be grown with care in the tropics but does not grow well as it is unsuited to tropical climate; prone to leaf-eating insects; prone to leaf spot (*Colletotrichum* species) or root rot (*Phytophthora* species) fungal diseases.

Plant Description:

- A short, aromatic, evergreen tree or shrub 2–15 m (6.6–49.2 ft) tall.
- The alternate, stalked, simple leaves have narrowly oval, leathery leaf blades with a finely serrated and undulate margins.
- Plants are unisexual, with female and male flowers on separate plants. Green-yellow flowers occur in axillary clusters of 4–5, with each about 1 cm (0.4 in) across.
- When ripe, the fruits are dark violet to glossy black, round to ellipsoid, and 1–2 cm (0.4–0.8 in) long.

Laurus nobilis leafy branch

Photos: (above and opposite middle) Hugh Tan Tiang Wah; (opposite top) Inna Shulman; (opposite bottom) Gypsy Flores

Laurus nobilis flowering branches

Saplings in a nursery

Flower buds with bushtit

Folklore, Beliefs, Uses and Other Interesting Facts

- In Greek mythology, the god Apollo's infatuation with Daphne and her abhorrence of his love were created by Eros (Cupid). To escape from Apollo, Daphne turned into a bay laurel tree. In Greek, Daphne means laurel. Apollo then decided to adorn himself with laurel leaves in remembrance of his lost love and thus the laurel tree became sacred to Apollo.
- In medieval Greece, successful scholars were given a wreath made of bay laurel leaves to wear on their heads—the *bacca laurea*. The words 'laureate' and 'baccalaureate', which imply achievement and fame, are therefore rooted in the Greek tradition. The laurel wreath was associated with nobility and crowned the heads of Roman emperors. In the Renaissance, it came also to be associated with moral virtue.
- In New Forest, South England, the laurel was planted for protection against all kinds of evil and had the power to fend off the devil and witches, as well as to protect one from thunder, lightning and forest fires.
- In the language of flowers, the laurel flower represents glory, triumph or the reward of merit, and the laurel leaf represents assured happiness.
- Based on feng shui beliefs, the dark green leafed and bushy laurel, representing wood, would be best planted in the east, southeast and south of the garden. It is inauspicious in the southwest or northeast and neutral for other zones.
- The name 'Laura' comes from the laurel.
- Its aromatic leaves are used as an important seasoning in European cuisines.
- Cultivars of laurel are grown as ornamental plants which may be sheared into various shapes.
- The oil, extracts or decoction is used medicinally as an astringent, narcotic, stimulant and stomachic, and for treating urinary organ problems, dropsy and rheumatic pains.

Lilium species & hybrids garden lily • lily

Scientific Species Name:
***Lilium* species** and **hybrids**
(Latin *lilium*, derived from the Greek *leirion*, lily)

Common Species Name:
garden lily, lily (English); *lis*, *lys* (French); *Lilie* (German); *kamala*, *kumuda*, *nalini*, *naragisa* (Hindi); *bakung* (Malay); *bǎi hé* [百合] (Mandarin); *lirio* (Spanish)

Scientific Family Name:
Liliaceae

Common Family Name:
lily family

Natural Distribution:
Temperate northern hemisphere southwards to the mountains of the Asian tropics in India and the Philippines

USDA 2–8 (AUS 1–2)

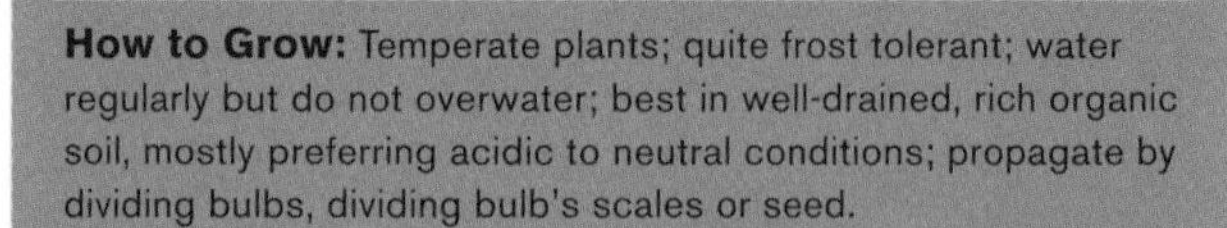

How to Grow: Temperate plants; quite frost tolerant; water regularly but do not overwater; best in well-drained, rich organic soil, mostly preferring acidic to neutral conditions; propagate by dividing bulbs, dividing bulb's scales or seed.

Drawbacks: Require a cold 'rest' period, so in the warmer areas, requires 4–6 weeks of refrigeration to simulate winter conditions; quite intolerant of waterlogged conditions, even for a few hours; seed-grown lilies may take up to 4 years to bloom.

Plant Description:

- Most lilies grown today are cultivated varieties of the thousands of hybrids (there are about 6,000 registered hybrids) and about 100 species.
- Perennial herbs that grow from bulbs.
- The erect, unbranched and leafy stems can grow to 3.1 m (10.2 ft) tall and are able to produce roots at its base and/or develop bulbils in the leaf axils.
- The whorled or alternate, simple, foliage leaves possess linear to elliptic leaf blades.
- The often fragrant flowers are usually found in clusters at the stem tips and consist of 6 tepals, 6 stamens and a pistil in the centre. The tepals occur in all colours except blue. They may be bell-, bowl- or trumpet-shaped with or without the tepal tips curved backwards to the flower stalk.
- The fruits are capsules which are round to elongated, splitting when ripe to release numerous, tiny seeds.

Lilium cultivar flower

 Photos: (above and opposite) Hugh Tan Tiang Wah

Flowers and buds on a leafy shoot of a *Lilium* cultivar

Upper and lower sides of *Lilium* cultivar leaf

Folklore, Beliefs, Uses and Other Interesting Facts

- The Madonna lily (*Lilium candidum*) is one of the most common symbols of the Virgin Mary. It is often depicted as being held by or growing at the side of Mary in graphical representations of the Annunciation.
- In the British Isles, the lily symbolises innocence, purity and virginity, hence it is the flower for weddings. To dream of lilies is a good omen. A man is believed to extinguish his womenfolk's purity by stepping on a lily plant. The lily is also used for funerals as it signifies a shriven soul untainted by sin. However, because of their funeral association, many dislike having the, in the home and some consider their presence a sign of impending death. Lily plants in a house garden will keep the house ghost-free as they are thought to repel ghosts.
- Based on feng shui beliefs, lilies are best grown in the sectors of the garden where their flower colour is most suitable. Colours like red, yellow and orange are preferred for their brightness, which attracts *yáng* energy. Red-, pink- or magenta-flowered cultivars are auspicious if grown at the left of the house front (because red is very auspicious, being the colour of blood and, hence, life and vigour, and symbolises the crimson phoenix, one of the celestial creatures that are the hallmarks of classical feng shui), and also in the south, southwest or northeast zones, but are inauspicious for the west and northwest, and neutral for the other zones. Yellow- or orange-flowered cultivars are auspicious if planted in the southwest, west and northwest zones, inauspicious in the north, and neutral for the rest. White-flowered cultivars should not be planted in the house front (because white is associated with death and mourning), east and southeast; however, they are auspicious for the west, northwest and north, and neutral for the rest.
- In the language of flowers, the lily represents majesty or purity. A rose-coloured lily symbolises rarity; a white one, candour or purity; and a yellow one, inquietude.
- Lilies are common ornamental plants, grown for their beautiful flowers in temperate gardens and parks; their bulbs and cut flowers are some of the most popular items in the horticultural trade.
- In the British Isles, lilies are used in folk medicine as first-aid cures for boils or to discharge pus from inflamed toes and fingers, and as a poultice.
- For 2,000 years or more in China, lilies have been cultivated, eaten and used as medicine.

Mangifera indica Indian mango

Scientific Species Name:
Mangifera indica
(Indian mango, vernacular name for one *Mangifera* species; Latin *fero*, to bear; Latin *indica*, of India)

Common Species Name:
Indian mango, mango (English); *mangue* (French); *Indischer Mango*, *Mango* (German); *am*, *amra* (Hindi); *ampelam*, *mangga*, *mempelam* (Malay); *máng guǒ* [芒果] (Mandarin); mango (Spanish)

Scientific Family Name:
Anacardiaceae

Common Family Name:
cashew nut family

Natural Distribution:
A cultigen of unknown progenitor species; supposedly originated in the Indo-Myanmar region

USDA >10B (AUS >4)

How to Grow: Fast growing, subtropical to tropical tree; well-drained sand, clay or loam, acidic to alkaline soils (pH 5.5–7 preferred); young parts frost intolerant; moderately tolerant of salt aerosol; moderately drought tolerant; propagate by grafting or seed.

Drawbacks: Branch wood tends to be weak and breaks in strong winds; produces much litter beneath the crown; in wet climates, the fruit sets badly and tends to be attacked by insects; plant resins may cause allergies; main problems are scale insects followed by sooty mould (growing on the exuded sugars) and fruit flies; numerous fungal diseases include bitter rot of fruit (*Glomerella cingulata*) and anthracnose (*Colletotrichum gloeosporioides*) of the leaves and fruit.

Plant Description:

- An evergreen tree to 10–45 m (32.8–147.6 ft) tall, with a roundish, symmetrical crown.
- The alternate, stalked, simple leaves have slightly leathery, lance-shaped or narrowly elliptic leaf blades up to 50 cm (19.7 in) long, with wavy edges. Young leaves are reddish, then yellowish, then shiny, dark green when mature.
- The small, yellowish-white to light brown flowers are densely borne on branched inflorescences.
- The fruit is a drupe, somewhat like that of a peach or plum, 4–25 cm (1.6–9.8 in) long, and almost round to longish-oblong. Ripe fruits may be green, or turn yellow, orange or red.
- The single, large seed is found within the woody endocarp, the innermost layer of the fruit wall.

Mangifera indica cultivar fruit

 Photos: (above and opposite) Hugh Tan Tiang Wah

Mangifera indica tree

Inflorescence

Mango and young coconut leaves festooned at the Sri Krishnan temple, Singapore

Folklore, Beliefs, Uses and Other Interesting Facts

- The Indian mango tree has been considered sacred by Buddhists ever since the Buddha rested beneath a grove of such trees to which he was bestowed.
- The mango tree is also considered sacred by the Hindus because of its great religious significance. The plant is believed to be an incarnation of Prajapati, the Lord of all Creation. In a popular Indian Hindu folklore called Surya Bai, a Rajah fell in love with a beautiful maiden named Surya Bai. The Rajah's first wife was jealous of Surya Bai and killed her; she became a sunflower and was later burnt into ashes by the queen's men. From the ashes of the sunflower sprang up a large mango tree which contained the spirit of Surya Bai. A single fruit ripened with her spirit dropped into the milk can of her long-lost mother. From this mango grew Surya Bai and, not long after, she was finally reunited with her family and the Rajah, who made her his Queen. The mango tree, therefore, is considered a tree of love by the Indians and is often prayed to for love wishes.
- Indian villagers believe that when a mango tree produces a flush of new leaves, a son is born, so fresh mango leaves are hung over the door of a house where there is a birth of a son. The mango plant is considered auspicious and its leaves are hung over the front door of a house where a marriage is to be conducted to bless the house and in the hope that the couple will produce a son. Mango leaves are also used in wedding ceremonies to bless the newly married couple with many children.
- The Indians also believe that mango wood is sacred, and include it in funeral pyres and in the *homa* (a ritual where offerings are made into a fire). Mango flowers are offered on the second day of Magh (the 11th month of the Hindu calendar) to the moon, to which they are dedicated, and also to Madan, the God of Love.
- Based on feng shui principles, the dense, leafy crown of the mango tree signifies the wood element, but the round crown signifies the opposing element, metal, which cuts wood. It may be best planted in the metal-element enhancing zones such as the west, northwest or north, but it is inauspicious for the east and southeast zones, and neutral for the rest.
- Numerous cultivars are primarily cultivated for their fruits, which are eaten unripe (Thailand and the Philippines), ripe (worldwide) or processed into chutneys, canned or dried slices, juices, pastes, pickles or purees. Mangoes are also grown in gardens as ornamental fruit trees. Mangoes have been introduced into the tropics and subtropics and have become invasive weeds in places such as Fiji, Guam and Hawaii.

Michelia ×*alba*

white champaca • white sandalwood

Scientific Species Name:
Michelia* ×*alba
(= *Magnolia* ×*alba*, *Michelia longifolia*)
(*Michelia* is named after Pietro Antonio Micheli [1671–1737], Florentine botanist; × = multiplication sign, denoting a hybrid status; Latin, white, referring to the white tepals of the flower of this hybrid)

Common Species Name:
white champaca, white chempaka, white sandalwood (English); *cempaka* [*chempaka*] *putih* [*puteh*], *cempaka* [*chempaka*] *gading* (Malay); *bái yù lán* [白玉兰] (Mandarin)

Scientific Family Name:
Magnoliaceae

Common Family Name:
magnolia family

Natural Distribution:
Hybrid of two *Michelia* species; now widely cultivated in the tropics and subtropics

USDA ≥10 (AUS ≥4)

How to Grow: Tropical tree; frost intolerant; well-drained, slightly acidic [pH 6.0–6.5], rich organic soil; propagation by seed or air-layering.

Drawbacks: Prone to scale insect infestation; fragrance of flowers may become overpowering; not for the superstitious!

Plant Description:

- Similar to *Michelia champaca* except for the cream instead of orange tepals of the flower. Also, this plant does not usually produce fruits, being a hybrid and more prone to being sterile.

 Photos: (opposite top) Giam Xingli; (opposite middle and bottom) Boo Chih Min

Michelia ×alba flower and buds on a leafy twig

Tree

Freshly opened flower

Folklore, Beliefs, Uses and Other Interesting Facts

- Cultivated as an ornamental tree of streets, parks or gardens for its beautiful, scented flowers, which are also used as religious altar offerings by the Chinese and Indians.
- In Malay folklore of Southeast Asia, trees with strongly fragrant flowers (especially at night) are often associated with ghosts or the *pontianak* as it is associated with a sickly sweet scent. Children are often warned against walking near such trees at night to avoid harm from these beings. To further enhance this reputation, the champaca is sometimes planted at Malay cemeteries.
- The fresh flowers are sold in Asian markets to be made into garlands, adornments for the hair of women and placed with clothes to perfume them.
- A dark green, extremely sweet, almost nauseatingly scented essential oil, distilled from the flowers, is used in perfumes. A fragrant essential oil is also distilled from the leaves.
- The timber is excellent.
- In Malay medicinal folklore, an infusion of the flower buds is given to women as a remedy for blood poisoning after a miscarriage.

Michelia champaca

champaka • orange chempaka

Scientific Species Name:
Michelia champaca
(= *Magnolia champaca*) (*Michelia* is named after Pietro Antonio Micheli [1671–1737], Florentine botanist; Hindu *champaca*, this species)

Common Species Name:
champa, champac, champaca, champak, champaka, fragrant champaca, fragrant Himalayan champaca, golden champaca, michelia, orange chempaka, sapu, yellow jade orchid tree (English); *cempa* [*chempa*], *cempaka* [*chempaka*], *cempaka* [*chempaka*] *merah*, *jampaka* (Malay); *huáng lán* [黄兰], *huáng yù lán* [黄玉兰] (Mandarin)

Scientific Family Name:
Magnoliaceae

Common Family Name:
magnolia family

Natural Distribution:
India, Indochina, Southwest China, Sumatra, Peninsular Malaysia, Java and the Lesser Sunda Islands

USDA ≥10 (AUS ≥4)

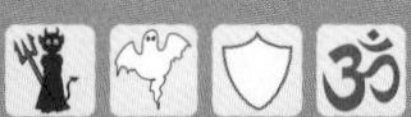

How to Grow: Tropical tree; frost intolerant; well-drained, slightly acidic [pH 6.0–6.5], rich organic soil; propagation by seed or air-layering.

Drawbacks: Prone to scale; attracts birds when fruiting; fragrance of flowers may become overpowering; not for the superstitious!

Plant Description:

- An evergreen, slow-growing tree growing to 50 m (164 ft) tall. Young trees have a conical crown. When older, branches of the tree are characteristically widely ascending, each bearing a conical crown.
- The stalked, spirally arranged leaves have yellowish green, oval to lance-shaped leaf blades, 10–30 cm (3.9–11.8 in) by 3.5–11 cm (1.4–4.3 in).
- The fragrant flowers usually occur singly at the leaf axils. Each consists of 15 light yellow becoming orange, elongated tepals surrounding numerous stamens, which surround the ovaries which are raised on a stalk.
- Each flower produces a grape-like bunch of woody fruits, which split when ripe to reveal thin, shiny black seeds that are covered in pink flesh.

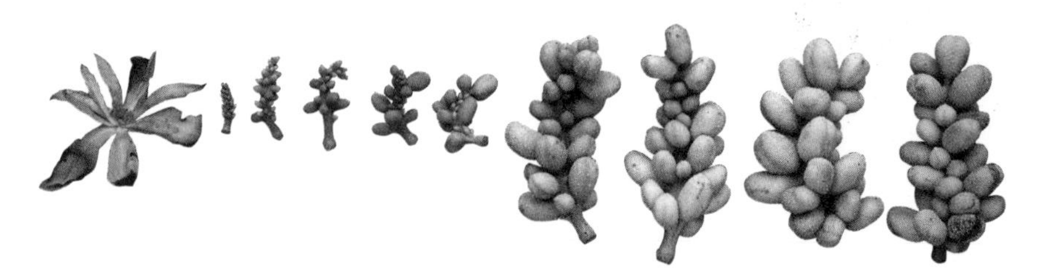

Michelia champaca fruit development from a single flower

 Photos: (above and opposite top) Hugh Tan Tiang Wah; (opposite middle) Boo Chih Min; (opposite bottom) Yeo Chow Khoon

Young *Michelia champaca* tree with its conical crown

Flower with some tepals removed to show the stamens surrounding the pistil

Fruits of a single flower

Folklore, Beliefs, Uses and Other Interesting Facts

- Cultivated around Hindu and Jain temples in India for the scent, which is associated with their gods. The trees provide fragrant blossoms for religious offerings here as well as in other Asian countries. Also cultivated as an ornamental street, park or garden tree for its beautiful, scented flowers.
- In Malay folklore of Southeast Asia, trees with strongly fragrant flowers (especially at night) are often associated with ghosts or the *pontianak*. Children are often warned against walking near such trees at night to avoid harm from these beings. To further enhance this reputation, the champaca is sometimes planted at Malay cemeteries.
- In the burial rites of the Malays, scented water is sprinkled over a new grave and the remainder on adjacent graves. Sandalwood water (*air gaharu cendana* in Malay) is prepared by mixing the finely shredded champaca flowers with other flowers and woods and soaking them in water.
- According to Malay folklore, a woman is particularly vulnerable to evil spirits during and immediately after childbirth. She is protected using pounded champaca leaves applied on the body as well as by drinking an infusion of champaca leaves and bark.
- The fresh flowers are sold in Asian markets to be made into garlands and adornments for women's hair. They are also placed in between clothes and sheets or in closets or handbags to perfume them. A Malay tradition is to sprinkle flower-soaked water onto the bed, clothes or hair to scent them.
- An essential oil, distilled from the flowers, is used in perfumes.
- The excellent timber is used to make canoes, carvings, furniture, tea boxes, windows and musical instruments. It is also used in house-building, panelling, planking or as fuel.
- The leaves are used for feeding silkworms.
- The bark decoction is used as an antipyretic in tropical Asia. The leaf juice is used for deworming and mixed with honey to relieve colic. The roots are used as a purgative and the flowers as a carminative, diuretic, stimulant or tonic. In Malay medical folklore, the seeds are pounded seeds and used in a formulation for treating rheumatism.
- In Suzhou, China, the flowers are used to perfume tea.

Musa species & hybrids banana

Scientific Species Name:
***Musa* species** and **hybrids**
(*Musa* from the Arabic *mouz* or *moz*, banana; the edible (seedless) bananas are cultivated varieties of *Musa acuminata* or of the hybrids of *Musa acuminata* and *Musa balbisiana* = *Musa* ×*paradisiaca*)

Common Species Name:
banana (English); *banane* (French); *Banane* (German); *kelaa* (Hindi); *pisang* (Malay); *xiāng jiāo* [香蕉] (Mandarin); *banana*, *plátano* (Spanish)

Scientific Family Name:
Musaceae

Common Family Name:
banana family

Natural Distribution:
Indo-Malaysian region to Northern Australia

USDA 9b–11 (AUS 3–7)

How to Grow: Fast-growing tropical or near tropical, giant herbs; wide soil pH and type ranges but slightly acid soils preferred (> pH 5); low tolerance for drought as it requires a lot of water for growth; has low to no tolerance for salt, frost or strong winds; propagate by suckers.

Drawbacks: Not for the superstitious; in climates with high rainfall throughout the year, the fruit sets badly and tends to be attacked by insects; needs heavy fertilising for good growth and fruit production; pests include nematodes, black weevil, banana rust thrip, banana spider mite, banana silvering thrip, fruit flies; diseases include Sigatoka (leaf spot) disease, Black Sigatoka (black leaf spot) disease, Panama disease, Moko disease.

Plant Description:

- Large herb that can grow up to 8 m (26.2 ft) tall.
- Technically, not a tree as the erect pseudostem on which the leaves are inserted is made up of the sheaths of the leaves. However, with flowering, the stem elongates through the leaf sheaths to produce the inflorescence at the top of the plant.
- The alternate leaves each consists of a sheath, stalk and oblong leaf blade.
- The inflorescence is pendulous and subtended by purple-red bracts (modified leaves). It consists of hand-like clusters of flowers inserted spirally along an axis.
- The fruit, often referred to as the edible or dessert banana, is yellow in colour with white pulp inside the yellow skin.

Musa Pisang Mas inflorescence at the stem tip

 Photos: (above and opposite) Hugh Tan Tiang Wah

Banana plant

Banana plants placed at the entrance of the Sri Krishnan Temple, Singapore

Folklore, Beliefs, Uses and Other Interesting Facts

- Among the Chinese, especially in the Hokkien dialect group, the banana fruit is regarded as auspicious. The Hokkien common name for the banana, *kim cheo*, is a phonetic pun for 'golden object'. Gold is considered by the Chinese to be the colour of wealth and royalty because golden robes were worn by Chinese emperors.
- Owing to its connection with wealth and good fortune, banana fruits are often presented as votive offerings to deities during festivities and whole plants are commonly cultivated in private gardens.
- Based on feng shui beliefs, the large, green, leafy banana plant is symbolic of the wood element, and is most auspicious in the east, southeast and south zones of the garden, inauspicious for the southwest and northeast, and neutral for the others.
- However, to the Southeast Asian Malays and Peranakans (Chinese descendants of immigrants to Southeast Asia who have partially adopted Malay culture), the banana plant is often frowned upon as a haunted plant inhabited by spirits or the *pontianak*. To this day, Malay children are often told by elders to stay away from banana trees for fear of a *pontianak* attack.
- The reddish brown sap from the banana plant (the result of phenol oxidase activity, the same chemical reaction caused by this enzyme which causes cut apples to turn brown) may have led to the idea of the plant 'bleeding' when cut. A superstition is that if one were to hammer a nail into the trunk of the banana plant and attach a string from it to one's toe, one will be woken at night by the *pontianak* tugging on the string.
- The connection between banana plants and *pontianak* can almost be considered a popular culture in the Malay Archipelago, owing to the popular depictions and accounts of ghost stories or movies featuring these two subjects.
- The banana plant is auspicious to the Hindus who believe the plant to be the incarnation of both Parvati (the wife of Siva, the destroyer) and Lakshmi (the wife of Vishnu, the protector). The leaves are used for religious ceremonies because they are considered sacred, and whole plants are placed at the entrance of temples and marital homes. The plant is a symbol of fertility. A bride is given a banana fruit to assure her of a male child and women who wish to have male children worship the plant in the month of Kartik (October to November).

Nelumbo nucifera lotus

Scientific Species Name:
Nelumbo nucifera
(= *Nelumbo speciosa*, *Nelumbium speciosum*) (Sinhalese *nelumbo*, lotus; Latin *nucifera*, nut-bearing, referring to the seeds of the lotus fruit)

Common Species Name:
East Indian lotus, Egyptian lotus, Indian lotus, lotus, Oriental lotus, sacred bean of India, sacred lotus, sacred water lily (English); *fève d'Egypte*, *lotus des Indes*, *lotus du Nil*, *lotus Indien*, *lotus magnolia*, *lotus sacré*, *lotus sacré de l'Inde* (French); *Indischer Lotus* (German); *kamal*, *kanwal*, *kumuda*, *padma* (Hindi); *bunga padam*, *bunga telepok*, *padema*, *seroja*, *teratai* (Malay); *hé* [荷], *lián* [蓮/莲] (Mandarin); *lotus*, *loto sagrado*, *rosa del Nilo* (Spanish)

Scientific Family Name:
Nelumbonaceae
(formerly in the *Nymphaeaceae*)

Common Family Name:
lotus family
(formerly in the water lily family)

Natural Distribution:
Lower Volga, South, Southeast and East Asia to Tropical Australia

USDA ≥4 (AUS ≥1)

How to Grow: Aquatic plant of stagnant or slow moving water; rhizomes should not freeze; rich loamy soil; grow in large watertight containers or in ponds; propagation by dividing rhizomes or scarifying seed then planting underwater.

Drawbacks: Needs fertilising regularly, or else the plant's leaves and flowers will grow smaller and smaller as nutrients are depleted, especially for potted plants; prone to scale and aphids sucking sap, or apple snails eating leaves.

Plant Description:

- Perennial aquatic plant which grows from a rhizome growing horizontally in the substrate up to 2 m (6.6 ft) deep underwater.
- Long-stalked leaves sprout from the rhizome and consist of both flexible leaf stalks with round, flat leaf blades floating on the water surface and leaf blades that are held stiffly above the water surface. Leaf blades can grow up to about 60 cm (23.6 in) or more across, are greyish green and covered with a waxy, water repellent surface.
- Flowers are fragrant and borne at the ends of a stiff erect stalk and held above the water surface. There are numerous pink, pink tinged or fading to white tepals surrounding numerous stamens surrounding the yellow, inverted cone-like receptacle which bear the carpels in surface cavities, each of which develops into fruits.
- The ovoid, 1–2 cm (0.4–0.8 in) long fruit, which is brown when ripe, contains one seed which can remain viable for hundreds of years!

Top view of *Nelumbo nucifera* receptacle showing unripe fruits in cavities

 Photos: (above, opposite middle and bottom) Hugh Tan Tiang Wah; (opposite top) Giam Xingli

Nelumbo nucifera flower and leaves in a pond

Double lotus flower with tepals folded

Side view of lotus bud with outer tepals folded

Folklore, Beliefs, Uses and Other Interesting Facts

- The lotus is considered sacred in China, India and Tibet, and is the source of the lotus motif of Asia. It symbolises eternity, plenty and good fortune.
- In Hindu mythology, the lotus is associated with Brahma, creator of the world. Brahma was borne by a lotus stalk that arose from the navel of Vishnu, the chief procreator of the universe. The lotus which bore Brahma is considered a duplicate manifestation of Padma, the wife of Vishnu, the lotus goddess.
- Chinese folk religion belief is that the cycle of blooming and wilting of the lotus flowers represents the perpetual cycles of reincarnation in Buddhism.
- In Vajrayana Buddhism belief, the ability of the lotus to grow out of the muddy water and flower beautifully is likened to the Vajrayana Buddhists trying to erase their mental defilements and achieve a clear state of mind, i.e., the 'Lotus Effect' derived from the minute papillae and water repellent wax on the upper leaf blade surface.
- A Chinese belief is that the lotus is a symbol of hardiness and success over obstacles despite difficult circumstances. This is because of its ability to survive and bloom despite its early growth in a muddy environment.
- According to feng shui beliefs, the lotus is the most auspicious flower one can grow because of its wonderful connotations: peace, hope, contentment and developing opportunities. The body of water that it grows in should be placed where the water element is favoured, i.e., the north, east and southeast sectors, not the south, and neutral for the others. Good growth of the plant produces growth of spiritual consciousness and fruiting signifies excellent good fortune.
- In the language of flowers, the lotus signifies eloquence, frivolity, silence or estranged love, and the lotus leaf, reincarnation.
- The lotus is the national flower of Vietnam and one of the two national flowers of Egypt, to which it was introduced in about 500 BC.
- The lotus flower's receptacle is the source of antihaemorrhagic quercitin. The fruits (or 'seeds' as they are known in commerce) are eaten by the Chinese in soups and desserts or are ground up as fillings in dumplings (after the bitter-tasting embryo is removed). The starchy rhizome, which can grow 20 m (65.6 ft) long in one growing season, is also eaten in soups or desserts in many East Asian and Southeast Asian countries.
- The red- and white-flowered scented cultivars of lotus are cultivated as an aquatic ornamental for pools in gardens or temples, where the double-flowered forms are frequently grown.

Nyctanthes arbor-tristis

night-blooming jasmine • tree of sorrow

Scientific Species Name:
Nyctanthes arbor-tristis
(Greek *nyx*, night, and *anthos*, a flower, referring to the night-blooming of this species; Latin *arbor*, a tree, and *tristis*, sad)

Common Species Name:
coral jasmine, night-blooming jasmine, night-flowering jasmine, night jasmine, sad tree, tree of sadness, tree of sorrow (English); *harsinghar*, *sephalika* (Hindi); *seri gading* (Malay); *yè huā* [夜花] (Mandarin)

Scientific Family Name:
Oleaceae

Common Family Name:
olive family

Natural Distribution:
Pakistan, Nepal, India and Thailand

USDA ≥11 (AUS ≥5)

How to Grow: Tropical tree or shrub; seasonal or aseasonal climates; well-drained soils; propagate by seed or stem cuttings.

Drawbacks: Untidy because very prone to insect attack; diseases include powdery mildew (leaves), leaf spot and others; not for the superstitious.

Plant Description:

- A shrub or small tree that grows up to 10 m (32.8 ft) tall.
- The opposite, hairy, shortly stalked leaves have leaf blades which are egg-shaped with smooth or slightly toothed margins, up to 12 cm (4.7 in) long, and inserted on twigs which are square in cross section.
- Flowers are arranged in axillary or terminal clusters of 2–7, subtended by purple-red bracts. The 5–8 petals, 2 cm (0.8 in) across, are joined together in the lower two-thirds into a bright orange tube while the rest are white, thus giving it a characteristic orange eye. The sweetly fragrant flowers bloom at sunset and wither after the following sunrise.
- The fruit is a heart-shaped to round, flattened, 2-cm (0.8 in) wide, dry capsule when ripe.
- One flattened seed is found in each of the 2 cells of the fruit.

 Photos: (opposite top) Boo Chih Min; (opposite middle) David Lee; (opposite bottom) Hugh Tan Tiang Wah

Nyctanthes arbor-tristis flowers and buds

Unripe fruit

Dried fruits on a twig

Folklore, Beliefs, Uses and Other Interesting Facts

- The tree of sadness is sacred to the Hindus, so it is often grown on the grounds of Hindu temples. Its name, 'tree of sorrow,' arose from Hindu mythology. A princess fell in love with Surya Deva, the handsome sun-god. But he deserted her and broke her heart, and so she commited suicide. She was cremated and this tree grew where her ashes fell. Hence, this plant is unable to tolerate the sun and blooms only at sunset, growing only in the shade of deep forests. At the first rays of light, the flowers (with orange centres) fall from the tree. The tree's drooping branches further enhances its name.
- In the *Skanda Purana* (one of the major eighteen *Puranas*, a Hindu religious text), the plant was said to be one of the products of *samudra manthan* (the churning of the ocean), an episodic event in Puranic mythology.
- The fragrant orange-white flowers are used as offerings of gratitude and devotion in the worship of Hindu gods. The presence of fragrance in Hindu worship is believed to please gods and lead to the fulfillment of wishes.
- A noncolourfast saffron yellow dye is extracted from the flowers of the plant. The flowers are collected from the floor in the morning. They are squeezed with water, and sometimes with the fruits, onto white cotton cloth.
- The plant has numerous medicinal uses. Its flowers can be used to induce menstruation and its leaves for treating rheumatism. The tree of sorrow is also reported for its analgesic, antihistaminic, anthelmintic, antipyretic, purgative, tranquilizing and ulcerogenic activity, amongst others.

Ocimum tenuiflorum holy basil • sacred basil

Scientific Species Name:
Ocimum tenuiflorum
(= *Ocimum sanctum*) (Greek *okimon*, an aromatic herb, possibly referring to an *Ocimum* species; Latin *tenuiflorus*, slender-flowered, referring to the slender inflorescences at the branch tips; Latin *sanctum*, holy, referring to the holy status of this plant to the Hindus)

Common Species Name:
holy basil, monks' basil, red basil, rough basil, sacred basil, sacred Thai basil, Siamese basil, Thai basil (English); *basilic sacré*, *basilic sacré à feuilles vertes*, *basilic Thaïlandais* (French); *Indisches Basilikum* (German); *baranda*, *jangalii tulasii*, *kaalaa tulasii*, *tulsi*, *Krishna tulasii*, *tulasii*, *varanda* (Hindi); *kemangi*, *selasih merah*, *selasih Siam* (Malay); *shèng luó lè* [圣罗勒/圣羅勒] (Mandarin)

Scientific Family Name:
Lamiaceae or **Labiatae**

Common Family Name:
mint family

Natural Distribution:
Old World tropics (e.g., India, Sri Lanka, Malaysia, Philippines, Indonesia, etc.); widely cultivated and also occurring as a weed of cultivation or wasteland

USDA >11 (AUS >5)

How to Grow: A frost-intolerant, tropical annual herb; well-drained soil, but well-watered, and mulched in pots or the ground; propagate from seed.

Drawbacks: Rather hardy, with little disease or pests; leaf miners attack the leaves occasionally; usually wanes after fruiting, but can be revived by severe pruning to produce new branches then prune again after flowering, and so on, to make this plant a perennial.

Plant Description:
- An aromatic woody or shrub-like herb up to 1 m (3.3 ft) tall.
- Stems and leaves are whitish and hairy.
- The opposite, stalked leaves have broadly oval leaf blades and up to 3 cm (1.2 in) long, with wavy and toothed margins.
- The small flowers are purplish or white and arranged in an elongated, loose inflorescence at the stem or branch tips.
- The fruit when ripe is dry and splits open into four one-seeded parts.

Ocimum tenuiflorum upper and lower sides of leaves

 Photos: (above and opposite) Hugh Tan Tiang Wah

Inflorescences at stem tip

Flowers and young fruits at inflorescence tip

Plants for sale in a nursery

Folklore, Beliefs, Uses and Other Interesting Facts

- Holy basil is a sacred plant of Hinduism and Vishnuism and found in many homes and temples in India.
- Holy basil is linked to the Hindu god, Vishnu the Preserver, one of the Trimurti, the triad of gods. The story of Vishnu defeating and killing the evil demon Jalandhar is one of many stories that illustrate the sacredness of holy basil. Jalandhar became a demon of immense power owing to the unyielding fidelity of his beautiful wife, Vrinda. He was invincible until Vishnu successfully tricked Vrinda into infidelity so Jalandhar's demonic power was reduced and he was defeated and killed by Vishnu. Vrinda was distraught by the death of her husband and killed herself in the funeral pyre. However, Vishnu's love for Vrinda prompted him to reincarnate her as the holy basil plant, worshipped by Hindu women today.
- As a tangible representation of the goddess Tulsi (Tulasi), the holy basil plant became an object of worship. The goddess Tulsi is also identified as Lakshmi (Vishnu's wife), Sita (wife of Rama, an incarnation of Vishnu) or Rukmini (wife of Krishna, an incarnation of Vishnu). Thus, it is believed that anyone cultivating the plant would obtain the love, blessings and protection of Vishnu.
- It is believed that by practising a daily ritual of prayer towards the holy basil, it can protect the family against harm by invoking the blessings of Vishnu and the goddess Tulsi.
- Holy basil is used in a ritual for a dying Brahmin to assure him that he will go straight to heaven.
- A holy basil twig offering to Vishnu in the month of Kartika (between October and November) is considered better than a thousand cows.
- A basil twig soaked in saffron offered to Vishnu is believed to make one like Vishnu, including his happiness.
- A basil twig offered to someone anxious and worrying is thought to eradicate his difficulties.
- Looking at holy basil pardons one from all of one's sins, touching it purifies one, and worshipping it cures one of all sins and brings happiness and wealth to the devotee.
- Holy basil is believed to be a demon killer and planted in homes to protect its inhabitants from evil spirits.
- The leaves of holy basil are not to be plucked on Sundays and Tuesdays or boiled, as these acts are believed to torture the plant's soul.

Olea europaea olive

Scientific Species Name:
Olea europaea
(Latin *olea*, this species; Latin *europaea*, European, referring to the natural distribution of this species)

Common Species Name:
olive (English); *olive* (French); *Olive* (German); *jaituna, jalapai, zaitun* (Hindi); *zait, zaitun* (Malay); *gǎn lǎn* [橄榄/橄欖] (Mandarin); *aceituna, oliva* (Spanish)

Scientific Family Name:
Oleaceae

Common Family Name:
olive family

Natural Distribution:
From the North Atlantic island archipelagos to Sind (a province in southeastern Pakistan), the Himalayas and South Africa

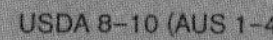

USDA 8–10 (AUS 1–4)

How to Grow: A long-lived, slow-growing, medium maintenance tree; fertile, medium moisture, well-drained soils; best in Mediterranean climates; slightly frost tolerant.

Drawbacks: No serious problems with pest or diseases; diseases include leaf or peacock spot, sooty mould, olive knot, verticillium wilt, bacterial canker and root rot; pests include olive fly (fruit), olive moth or kernel borer (fruit), black scale (branch and trunk), psyllids (flowers), mites (fruit and leaf), thrips (flower and young leaf) and scale insects (leaf and twigs); can create a mess under the tree during fruiting; can be grown in the tropics but does not flower or fruit.

Plant Description:

- An evergreen, strongly branched shrub growing to 5 m (16.4 ft) tall or a tree growing to 20 m (65.6 ft) tall, with an often fluted or crooked trunk. Young branches are whitish and thorny.
- The stalked, opposite leaves each have an elliptic or lance-shaped, leathery leaf blade which is dark grey-green above, densely scaly and silvery below, and 3–9 cm (1.2–3.5 in) long and 0.5–3 cm (0.2–1.2 in) wide.
- The tiny, fragrant, white-petalled, bisexual flowers are borne in axillary, branching, many-flowered inflorescences 3–8 cm (1.2–3.1 in) long.
- The fruit is a round to ellipsoid drupe 0.5–6 cm (0.2–2.4 in) long and 0.5–2.5 cm (0.2–1.0 in) wide, bright green, turning purple-black, brown-green or ivory-white when ripe. The middle fruit wall layer (mesocarp) is oil-rich, surrounding the stone (pyrene) which consists of the hard innermost fruit wall layer, which envelops usually 1 ellipsoid seed, 9–11 mm (0.35–0.43 in) long.

Olea europaea leafy branch

 Photos: (above and opposite bottom) Hugh Tan Tiang Wah; (opposite top) Nicole Hamaker; (opposite middle) Vicky Scott

Tree in Nicosia, Sicily

Olea europaea fruiting branches

Pickled olives

Folklore, Beliefs, Uses and Other Interesting Facts

- Since pre-Christian times, an olive branch (foliage) was the symbol of peace and goodwill because its oil could cure human ills and calm troubles. The saying "to hold out the/an olive branch," meaning to make peace, reflects this.
- Greek mythology has it that the olive was the Goddess Athena's gift to humans. To commemorate this, Athens was named in her honour. In ancient Greece, the greatest honour that could be bestowed on a citizen was a wreath of the wild olive, which was also an ancient Olympic prize. It should be noted that the Olympic Games was a time of truce amongst the warring city states of Greece, so an olive wreath was most appropriate.
- In Deuteronomy 8:7–8 (King James Version), the people of Israel were promised "....a good land....A land of wheat, and barley, and vines, and fig trees, and pomegranates; a land of olive oil, and honey", so these have a special place in Judaism as the seven plants of the Bible.
- The olive is also one of the most commonly cultivated plants of the Bible lands. 'Olive' is mentioned at least 25 times in the Bible (King James Version).
- A dove brings back an olive leaf to Noah to indicate the subsidence of the great flood (Genesis 8:11).
- The olive is used as a symbol of prosperity and vigour: "Your wife will be like a fruitful vine within your house; your sons will be like olive shoots around your table." (Psalm 128:3, King James Version).
- The olive is also mentioned six times in the Koran, one of the fewer than 20 plant species mentioned: "I swear by the fig and the olive." (95:1, M.H. Shakir translation).
- Based on feng shui beliefs, the olive tree with its silvery leaves would represent the metal element, and its bushiness, prosperity; so it is auspicious if planted in the west, northwest and north zones of the garden, but inauspicious if planted in the east and southeast zones, and neutral for the rest.
- In the language of flowers, the olive represents peace and the olive tree, charity.
- As a gesture of goodwill, the soldiers of the 164th Corps Support Group of the United States Armed Forces planted olive trees in Iraq, the first three in a ceremony on 23 December 2006, and the remainder a few days later to total 164 plants in all. The olive is a suitable choice as it is a symbol of peace and also well-suited to the climate of Iraq. The plants were donated by the California Rare Fruit Growers, Incorporated. American gardening enthusiasts held a parallel ceremony on the same day in San Luis Obispo, California, to symbolise this gift of peace from the American people to the Iraqis.
- Olive trees are cultivated in the Mediterranean, Australia, South Africa, Mexico, and California for their fruit. Olives are also preserved in brine as an appetiser, etc., and olive oil is used for cooking, in salads, tinned sardines, and for medicinal usage.

Oryza sativa rice

Scientific Species Name:
Oryza sativa
(Greek and Latin, *oryza*, rice; Latin *sativa*, cultivated)

Common Species Name:
Asian rice, common rice, cultivated rice, lowland rice, paddy rice, rice, upland rice (English); *riz, riz cargo, riz commun, riz cultivé, riz non décortiqué, riz de plaine, riz paddy, riz vêtu* (French); *Gemeiner Reis, Paddy-Reis, Reis, Rohreis* (German); *chavala, sela, dhana* (Hindi); *nasi, padi* (Malay); *dào zǐ* [稻子], *mǐ* [米], *shǔi dào* [水稻], *yà zhōu zāi péi dào* [亚洲栽培稻/亞洲栽培稻] (Mandarin); *arroz, arroz con cáscara, arroz con cáscara asiático, arroz irrigado* (Spanish)

Scientific Family Name:
Poaceae or **Gramineae**

Common Family Name:
grass family

Natural Distribution:
A possible derivative of *Oryza rufipogon* (brownbeard rice), with probably several centres of domestication from about 2,000 BC in the Lower Yangtze River valley, North India or Thailand)

USDA ≥6 (AUS ≥1)

How to Grow: Temperate to tropical, aquatic plant; needs to be flooded (lowland rice) or not (upland or dry rice); different cultivars have different tolerances for cold, salinity, etc., so should be selected for the specific locality; heavy fertile soils are best, pH 6.5–7.0; propagation by seed or tillering.

Drawbacks: Annual plant which dies after fruiting; affected by rice blast disease (fungal), bacterial leaf blight and tungro disease (viral); insect pests include the brown planthopper and striped borer; birds also eat the grains.

Plant Description:

- An erect annual herb which grows up to 50–180 cm (19.7–70.9 in) tall, or to 5 m (16.4 ft) long in deep water cultivars; with leaves arising along the ascending or erect stem.
- The alternate leaves, in two rows, consist of a sheath topped by a long, narrow blade up to 60 by 2.2 cm (23.6 by 0.9 in).
- The 50–500 tiny flowers are borne in a branching inflorescence, up to 45 cm (17.7 in) long, bending downwards from the tip.
- The fruits are grains, and their coverings mature from green to yellow or purplish-brown.

Spikelets, one with the stamens (yellow) and stigma (black) protruding out

 Photos: (above and opposite top) Hugh Tan Tiang Wah; (opposite bottom) Keith Weller, ARS, USDA

Oryza sativa infructescences bearing rice grains

Long grain US rice

Folklore, Beliefs, Uses and Other Interesting Facts

- For the Chinese, rice (*mǐ*) represents prosperity and an abundance of food. During the Chinese New Year, the rice plant is an auspicious ornamental plant which symbolises a bountiful harvest for the year ahead. If potted rice plants are unavailable, small bowls of rice are sometimes used to adorn the main dining table for the same auspicious reasons. Based on feng shui beliefs, this is one of the plants to place near the door of the house during Chinese New Year, despite its drooping shoots (heavy with ripe rice grains).
- As it is cultivated in Southeast Asia, it is not surprising that rice plays an important part in the folklore and the belief systems of the indigenous Southeast Asians. To the Malays, the rice plant is the subject of several proverbs, one of which is *'Jadilah orang pandai bagai padi yang merunduk'* ('Be a smart man like drooping rice'). The fruiting shoot droops more as it matures as the grains get heavier and heavier. The drooping rice shoot is compared to the drooping of the head, which implies humility. Rice (or *padi*) is therefore a sign of humility, which is highly regarded in Malay society.
- In Malay magic rituals, rice flour is often the base ingredient of the potion (which includes parts of other magic plants) that is spread on the afflicted to drive out evil spirits. Cooked rice is also used as an offering to appease spirits.
- Rice is the key source of carbohydrate in Asia and is probably the world's most important food crop, being the staple food of 40 per cent of humankind. The polished rice grains (with their husks removed) are eaten boiled or steamed with dishes; the flour is used in breakfast foods, baby foods, confectionaries, etc., and cosmetics; broken rice grains are used in foods, starch and textile manufacture; fermented to produce Chinese rice wine, Japanese sake, other rice-wines, rice-beer, rice-spirits, etc.; rice straw is utilised for animal fodder or bedding; husks used for mulching, fuel, bedding, as an absorbent, building board, carrier for vitamins, etc.; rice bran from polishing the grains is used for livestock and poultry feed and rice bran oils are used for manufacturing glycerine and soap.

Pachira aquatica money tree • knotty pachira

Scientific Species Name:
Pachira aquatica
(*Pachira*, from the Guyanese name of the tree; Latin *aquatica*, growing in or near water)

Common Species Name:
American chestnut, Guiana chestnut, knotty pachira, Malabar chestnut, money plant, money tree, provision tree, saba nut, water chestnut (English); *cacaoyer-rivière*, *châtaignier de la Guyane*, *chataigner sauvage*, *noisetier de la Guyane*, *pachirier aquatique* (French); *wilder Cacaobaum*, *wilder Kakaobaum* (German); *zhāo cái shù* [招财树/招財樹] (Mandarin); *cacao cimarrón*, *cacao de monte*, *cacao de playa*, *castaño*, *castaño de agua*, *castaño de la Guayana*, *castaño silvestre*, *ceibo de agua*, *ceibo de arroyo*, *chila blanca*, *jelinjoche*, *palo de boya tetón*, *pombo*, *pumpunjuche*, *quirihillo*, *sapotolón*, *sunzapote*, *tsine*, *zapote bobo*, *zapote de agua*, *zapotolongo*, *zapotón* (Spanish)

Scientific Family Name:
Malvaceae
(formerly in the Bombacaceae)

Common Family Name:
cotton family
(formerly in the silk cotton tree family)

Natural Distribution:
Mexico, Central America, Northern and Western South America and Northern Brazil; in freshwater swamps of tidal estuaries or river banks

USDA ≥10 (AUS ≥4)

How to Grow: Tropical tree; just tolerant of slight subzero temperatures for a short while; tolerant of flooding, so can grow at pond edges or in standing water; propagated by seed and stem cuttings.

Drawbacks: Generally hardy plant; potted plants often grow out of shape so must be pruned; prone to scale insects.

Plant Description:

- An evergreen tree growing to 18 m (59 ft) tall.
- The spirally arranged, stalked leaves are palmate with the leaflets arising from the tip of the leaf stalk. Each stalked leaflet is 15–30 cm (6.0–11.8 in) long, with a narrowly elliptic blade.
- The large, showy flowers are arranged in small clusters. Petals are narrow and cream or pink to purple, curling backwards from the numerous, 7.5–10 cm (3.0–3.9 in) long stamens whose filaments are white in the lower half and red in the upper half.
- The fruit is a woody capsule, somewhat round to elongate-elliptic and up to 30 cm (11.8 in) long, containing numerous, angular, woody seeds.

Photos: (opposite top) Giam Xingli; (opposite middle) Angie Ng-Chua; (opposite bottom) Hugh Tan Tiang Wah

Pachira aquatica plants with braided stems

Folklore, Beliefs, Uses and Other Interesting Facts

- The money tree is now a popular ornamental tree, favoured the world over for its auspiciousness. It is believed to be able to help its owner amass huge wealth.
- In 1986, Taiwanese truck driver Wang Ching-fu braided the stems of five individual young plants together. Since then, braided stemmed money trees have became very popular and such plants sell very well. The saying 'The five fortunes come home, richer at each juncture' has come to describe such plants. It is further believed that the braided stems act as an obstacle to trap wealth and prevent the outflow of money from the home or business. In 2005, Taiwan exported NT$ 250 million (US$ 7.6 million) worth of money trees, and over two decades, NT$ 2 to 3 billion (US$ 0.62 billion to US$ 0.93 billion). These plants are also commercially cultivated in China, Thailand and Vietnam.
- During Chinese New Year, it is common for the Chinese to tie gold and/or red ribbons on the branches of the money tree to enhance the auspiciousness of this plant. Two pots of trees are placed at both sides of the front door to invite good fortune and prosperity.
- The seeds, from the split fruit, taste like peanuts when fresh but taste like chestnuts when roasted or fried in oil. The seed can also be ground into a bread flour. Young leaves and flowers are eaten as vegetables. The trunks are a source of lightweight hardwood timber and also used for paper pulp.

Flowers and leaves

Fruits and leaves

Paeonia cultivars peony

Scientific Species Name:
***Paeonia* cultivars**
(Greek *paeonia*, peony; 33 species)

Common Species Name:
paeony, peony (English); *pivoine* (French); *Päonie* (German); *mŭ dān* [牡丹], *sháo yào* [芍药] (Mandarin); *peonía* (Spanish)

Scientific Family Name:
Paeoniaceae

Common Family Name:
peony family

Natural Distribution:
Temperate Europe, Asia and Western North America

USDA 2–8 (AUS 1–2)

How to Grow: Temperate plant; water regularly, but do not overwater; deep, rich, organic, slightly acidic to slightly alkaline soil; propagate by dividing the rhizomes.

Drawbacks: Bud development begins in late summer. Flowers form only after a cold spell (winter) but can be induced to flower out of season by treatment with gibberellic acid; fungal diseases include peony wilt disease; plants are attracted to the flowers by the nectars from their nectaries.

Plant Description:

- Hardy, perennial shrubs or herbs with rhizomes and tuberous roots. Shrubby peonies, commonly known as 'tree peonies', have a slender, cane-like stem that exhibits branching near the tip.
- The compound leaves occur both at the stem's base (basal leaves) and along the rest of the length (cauline leaves) in herbaceous species. Only cauline leaves, however, are found in the shrubby species.
- The fragrant flowers are large and showy. Each flower has 5 sepals and 5 to many red, pink, orange, yellow, purple or white petals. There are numerous stamens. Flowers may be 'single' (5 petals), 'double' (many petals) or 'anemone', with large outer petals and small inner ones clustered into a ball.
- Each flower produces 1–6 fruits that are dry when ripe, and splitting along a single suture to release the seeds.

 Photos: (opposite) Anna Buxton

Paeonia cultivar flowers on leafy branches

Paeonia cultivar flower and bud

Folklore, Beliefs, Uses and Other Interesting Facts

- The peony is seen by the Chinese as the king of flowers (*huā wáng* [花王]) owing to its large, majestic blooms. They also believe that the peony brings honour, affection and love, wealth and good luck, especially in romance, as well as signifying happiness and immortality. As it flowers in spring, the peony is considered one of the auspicious plants for decoration in the Chinese New Year, as it is a symbol of prosperity. It is considered a *yáng* (male) flower as well as a symbol of feminine beauty. It is considered an omen of extremely good fortune if a peony plant bursts into full bloom with vivid green foliage. Colourful paintings of peony plants in bloom are hung in homes so that the occupants can have this good fortune all the time!

- Based on feng shui beliefs, a peony planted in the southwest corner ('big earth' zone) of the garden will draw luck for good relationships, marriage and romance into one's home. The best is a red-flowered cultivated variety, although all the other colours are also auspicious, except for white. Red is considered the luckiest for families with daughters looking for husbands. White is the colour of mourning and funerals so white peony flowers are used to decorate coffins. For tropical gardens, peony substitutes include 'double flower' begonia (*Begonia*) or hibiscus (*Hibiscus*) cultivated varieties.

- The origin of the peony is explained in a couple of Greek myths. In one of them, Paeon, the physician of the Greek gods on Mount Olympus, was killed because of the jealousy of his teacher, Aesculapius. However, Hades, the god of the underworld, who was once cured of an arrow wound by Paeon, turned him into the peony plant.

- Another Greek myth has it that Leto, the mother of Apollo, informed Paeon of a herb on Mount Olympus which could be useful for easing childbirth. Paeon proceeded to use the herb and, thereafter, it became associated with Paeon.

- The ancient Greeks believed the peony's root was a powerful cure to counteract many types of evil and also a remedy for epilepsy, insanity, nightmares and similar mental afflictions. To be effective, the root had to be extracted from the ground by a dog tied to it, attracted to a piece of meat placed beyond its reach. As it emerged, the plant is thought to utter a scream that is fatal to its listeners, so humans had to keep their distance during this operation.

- Roy Vickery cited two recent reports from the British Isles about the belief that an odd number of flowers on a peony plant is a sign of death in the household where the plant is found. However, in West Sussex, England, a necklace made from the roots of the peony is worn by children to prevent convulsions and aid teething. The latter belief is likely derived from the belief in mid-19th century rural England that a seed necklace is worn as a charm to prevent the same. This, in turn, is likely derived from Pythagoras's observation, in the time of the ancient Greeks, that the root or seed necklace would protect the wearer from the recurrence of epilepsy. In the British Isles, peony seeds and roots, particularly those of the male plant, are considered to offer protection and, hence, used in anti-witch charms and amulets.

- In the language of flowers, the peony represents bashfulness, shame, bashful shame, hardiness, heaviness, anger or a frown.

- The peony (*Paeonia lactiflora*) is the state flower of Indiana, USA.

Phoenix dactylifera date palm

Scientific Species Name:
Phoenix dactylifera
(Greek *phoinix*, date palm; Latin *dactylifera*, fingerlike)

Common Species Name:
date, date palm, dry date, edible date, red date, semisoft date, soft date (English); *dattier, dattier commun, palmier dattier* (French); *Dattelpalme* (German); *khaji, khajur, salma, sendhi* (Hindi); *kurma, tamar* (Malay); *hǎi zǎo* [海枣/海棗], *yē zǎo* [椰枣/椰棗] (Mandarin); *dátil, palmera datilera, palmera de dátiles* (Spanish)

Scientific Family Name:
Arecaceae or **Palmae**

Common Family Name:
palm family

Natural Distribution:
A cultigen, although possibly wild in the Northeast Sahara and Arabia where trees with small inedible fruit are found

USDA 8–11 (AUS 2–5)

How to Grow: Slow-growing palm; well to medium-drained, well aerated, slightly acid to slightly alkaline, sand, sandy loam, clay and other heavy soils; high drought tolerance; medium soil salt tolerance; propagate by 3- to 5-year-old offshoots or suckers, or germinate by seed (keep seed and seedlings wet), but offspring may not be true to type and 50 per cent may be male.

Drawbacks: Has high wind resistance, so may be blown over in strong winds; spiny plant, so difficult to maintain; insect pests may attack unripe or ripe fruits, leaf bases (weevils), leaves and trunks (scale), apical meristems (beetles); birds and lizards may also pilfer the fruits; fungal diseases include the Bayoud disease, pinhead spot, grey blight, and spongy white rot, others cause leaf pustules, inflorescence decay, 'black scorch'; lethal yellowing viral disease.

Plant Description:

- A multiple-trunked palm that grows up to 40 m (131.2 ft) tall. A side shoot often emerges after 6–16 years and is removed to make it grow as a single-trunked tree.
- The spirally arranged, stalked, pinnate leaves grow to 6 m (19.7 ft) long. The leaf stalk is long and spiny, the midrib is stout and the leaflets are greyish or bluish green, 20–40 cm (7.9–15.7 in) long and folded lengthwise. The lowermost leaflets are spines.
- Most plants are male or female, but some bear both male and female flowers, or bisexual flowers. The small and slightly fragrant flowers are inserted on a much branched inflorescence of up to 6,000–10,000 flowers. Female flowers are whitish and found in inflorescences 30–75 cm (11.8–29.5 in) long, whilst the male flowers are cream-coloured and borne in inflorescences 15–22.5 cm (5.9–8.9 in) long.
- The fruit is an oblong drupe and, when ripe, dark brown, reddish or yellowish brown, and 2–8 cm (0.8–3.1 in) long. The 'skin' (outermost layer of the fruit wall) is usually thin and the 'flesh' (the middle layer of the fruit wall) soft and sweet (astringent until ripe; about 60–70 per cent sugar). There is a single grooved 'stone,' consisting of the hard last fruit wall layer surrounding the single seed.

Side view of *Phoenix dactylifera* dried fruit

 Photos: (above, opposite middle and bottom) Hugh Tan Tiang Wah; (opposite top) Jeff Vanuga, USDA Natural Resources Conservation Service

Date palms in Arizona, USA

'Stones' from the fruit

Dried fruit, with view of sepals

Folklore, Beliefs, Uses and Other Interesting Facts

- In the Middle East, where the date palm is native and possibly cultivated for about 4,000 years with numerous uses, it is still much revered and regarded as a symbol of fertility, depicted in bas-relief and on coins. Carvings of the Northwest Palace of the Assyrian King Ashurnasirpal II (883–859 BC) at Nimrud depict hand-pollination of the date palm.
- Physicians of the Sumer civilisation between the Tigris and Euphrates Rivers (one of the earliest civilisations; now Southern Iraq) utilised the roots, bark, leaves and seeds of the date palm. The panicles were used as amulets and magical weapons. Magician priests also used date panicles as magical weapons in banishing evil. The sap, tapped from the trunks of older date palms, is fermented into palm wine. Palm wine is used as a ritual inebriant and it was valued for its aphrodisiac properties.
- In Deuteronomy 8:7–8 (King James Version), the people of Israel were promised "....a good land....A land of wheat, and barley, and vines, and fig trees, and pomegranates; a land of oil olive, and honey," so these have a special place in Judaism as the seven plants of the Bible. The 'honey' mentioned above is interpreted as the fruit of the date palm. The date palm is also one of the most commonly cultivated plants of the Bible lands.
- In the rest of the Bible, Jericho is described as "the city of palm trees" (Deuteronomy 34:3). Palm leaves are used to welcome Jesus' triumphal entry to Jerusalem (John 12:13). Psalms 92:12 (King James Version) states: "The righteous shall flourish like the palm tree: he shall grow like a cedar in Lebanon." Here, the palm is a symbol of victory.
- The Bible also instructs Jews to celebrate the Festival of Sukkot, an eight-day festival of booths (temporary dwellings) and the autumn harvest, a time of thanksgiving for God's presence, by utilising the four species (or *arba minim* in Hebrew), citron (Hebrew *etrog*; *Citrus medica*), a date palm leaf (*lulav*), willow branch (*arava*) and myrtle branch (*hadas*): "And ye shall take you on the first day the fruit of goodly trees, branches of palm-trees, and boughs of thick trees, and willows of the brook, and ye shall rejoice before HaShem your G-d seven days." (Leviticus 23:40, Jewish Publication Society 1917 Version).
- The date palm is not auspicious for a garden because of its spiny characteristic. Based on feng shui beliefs, the plant is covered with poison arrows which emit hostile energy. Furthermore, the tall, narrow, single trunk resembles a large poison arrow.
- In the language of flowers, the palm symbolises dignity or victory.

Pinus species pine

Scientific Species Name:
***Pinus* species**
(Latin *pinus*, pine, particularly the parasol, stone or umbrella pine [*Pinus pinea*]; there are about 100 species of pine)

Common Species Name:
pine (English); *pignon*, *pin* (French); *Kiefer* (German); *pain* (Malay); *sōng* [松] (Mandarin); *pino* (Spanish)

Scientific Family Name:
Pinaceae

Common Family Name:
pine family

Natural Distribution:
Northern temperate regions, from the Arctic Circle south to Central America, North Africa, Sumatra and Java

How to Grow: Most are temperate species, mostly preferring well-drained, acidic soils, some are fire and drought-tolerant; some species like the Caribbean pine (*Pinus caribaea*) can grow in the lowland tropics; propagation is mostly by seed (which have a long viability period in storage), less often by stem cuttings or tissue culture.

Drawbacks: Some species may become invasive weeds in their country of introduction; refer specifically to separate references for pests and diseases for individual species.

Plant Description:

- Pines are evergreen, resinous trees or shrubs.
- The crown is conical when young but becomes more rounded or flat-topped with age.
- Needle and scale leaves are present in adult trees. Needle leaves are slender, elongated, 1–7 (usually 2–5) per bundle, growing from a short shoot. There are also scale leaves, which grow on a long shoot from which the short shoots may sprout, one each in the axils of the scale leaves. Branches grow in pseudowhorls (very tight spirals) at the long shoot's tip.
- Plants bear male and female cones on the same branch. Male cones are oval to cylindrical, yellow, red, orange or lavender, and produced in spirally arranged clusters. Female cones are oval-conical, solitary or in clusters. Scales on female cones are slightly to very woody. The female cone takes 2–3 years to develop after wind-pollination.
- Two winged seeds are produced from each female cone scale.

Pinus sylvestris seeds

 Photos: (above) Steve Hurst @ USDA-NRCS PLANTS Database; (opposite) Hugh Tan Tiang Wah

Pinus caribaea cones on branches with needle leaves

Pinus caribaea tree

Pinus caribaea trunk bark

Folklore, Beliefs, Uses and Other Interesting Facts

- In an ancient Greek legend, set in the forests of Greece, the nymph Pitys was in charge of tending pine trees. She had a lover, Boreas, god of the north wind, but still flirted with Pan, a minor male god. Boreas found out about it and, in the quarrel that ensued, threw Pitys against a rocky ledge. She immediately turned into a pine tree. The resin droplets on wounded twigs are said to be her tears.

- Pines are considered unlucky trees in the British crown dependency of Guernsey. There, it is thought that if one planted a row of pines, the property on which they grew would change ownership. Also, the one who falls asleep beneath a pine tree will die.

- In Silesia (a region now divided by the borders of Poland, Germany and Czech Republic), rosin (refer to the final paragraph) was burned at night between Christmas and New Year so that the smoke can drive witches and evil spirits from the house.

- In traditional Chinese culture, the pine, the bamboo (Bambuseae) and the plum blossom (*Prunus ×domestica*) are the 'Three Friends of the Cold Season'. The pine is a symbol of longevity, possibly because it is evergreen and stays green and leafy throughout the year. Planting a pine adjacent to a cypress tree (*Cupressus* species) symbolises eternal friendship through hardships, as both stay evergreen through the snows of winter. Based on feng shui beliefs, plant a pine close to the back of the house (for support) but prune it to a desired height and to suit the size and shape of one's garden.

- In the language of flowers, the pine represents hardiness, light or pity. The pitch pine (*Pinus palustris*) signifies 'You are perfect'.

- The resin tapped from pines is further refined into rosin and turpentine oil. Turpentine is commonly used in households as a paint or varnish solvent or thinner. It also has medicinal properties and can be used as a stimulant, antispasmodic, astringent, diuretic and anti-pathogenic. Pine wood is one of the most commercially important of temperate timbers.

Platycladus orientalis

Chinese *arbor-vitae* • oriental thuja

Scientific Species Name:
Platycladus orientalis
(= *Biota orientalis*, *Thuja orientalis*) (Greek *platy-*, flat; Greek *kladion*, branch, referring to the flat branches; Latin *orientalis*, Eastern or from the Orient; referring to its natural distribution; Latin *biota*, life; Greek *thuia*, a kind of juniper [*Juniperus* species])

Common Species Name:
oriental thuja, Chinese *arbor-vitae*, Oriental *arbor-vitae* (English); *thuya le Chine* (French); *Morgenländische Lebensbaum* (German); *cè bǎi* [侧柏/側柏] (Mandarin); *arbor-vitae chino*, *biota* (Spanish)

Scientific Family Name:
Cupressaceae

Common Family Name:
cypress family

Natural Distribution:
Central China (Southern Gansu, Hebei, Henan, Shaanxi, Shanxi). Extensively cultivated in China and elsewhere.

USDA 6–9 (AUS 1–3)

How to Grow: A slow-growing, medium-maintenance, temperate, evergreen tree that can even grow in the tropics (but will not produce cones); can be trained into a hedge or large shrub; fertile, medium moisture, well-drained soils; shelter from strong winds; not tolerant of excessive frost; propagate by cuttings and seed.

Drawbacks: Generally pest- and disease-free when well-maintained; crowns open with age and appear more untidy; pests include bag worms and spider mites; branches may break in winter from the weight of the snow.

Plant Description:

- An evergreen, coniferous tree growing to about 20 m (65.6 ft) tall. The reddish to light grey-brown trunk bark is thin and long-flaking. Dwarf cultivars are commonly used in landscaping.
- The crown of a young tree is ovoid-conical but broadly round or irregular when old. Branches are flat sprays that are perpendicular to the crown's surface.
- The opposite, densely arranged, 4-ranked, scale leaves are 1–3 mm (0.04–0.12 in) long, and grow flat on the stem. They have a distinct, longitudinal, glandular furrow below.
- Each tree bears male and female cones. Male (pollen-bearing) cones are egg-shaped, yellowish green, 2–3 mm (0.08–0.12 in) long. Female (seed-bearing) cones are somewhat round and bluish green when unripe, somewhat egg-shaped and reddish brown, 1.5–2.5 cm (0.6–1.0 in) long by 1–1.8 cm (0.4–0.7 in) wide when ripe. Only the middle 2 pairs of cone scales can bear seeds; the basal 2 cone scales are 2-seeded, and the apical 2 cone scales, one-seeded.
- The wingless seed is grey-brown or purple-brown, egg-shaped or somewhat ellipsoid, 5–7 mm (0.20–0.28 in) long by 3–4 mm (0.12–0.16 in) wide.

 Photos: (opposite top) David Lee; (opposite middle) Boo Chih Min; (opposite bottom) Hugh Tan Tiang Wah

Mature female cones of *Platycladus orientalis*

Trees in Singapore

Platycadus orientalis cultivar potted sapling

Folklore, Beliefs, Uses and Other Interesting Facts

- The name *arbor-vitae* or *arborvitae* (Latin for 'tree of life') applies to *Platycladus* or *Thuja* species. *Platycladus orientalis* (Chinese *arbor-vitae*) is the only species in *Platycladus*. *Platycladus orientalis* and *Thuja occidentalis* are the two most commonly cultivated *arbor-vitae*.
- Some Chinese *arbor-vitae* trees in Beijing are reported to be older than 1,000 years! A tree in the Forbidden City is said to be the 'life tree of the Manchu Dynasty,' and so was never pruned or cut because it was believed that would affect the power and status of the Manchus, who identified with that tree. When old, it was even propped up to prevent it from falling down.
- The *arbor-vitae* species is regarded as a symbol of stability, longevity or immortality in Chinese and Japanese tradition, and frequently planted at shrines and temples.
- The Chinese of the Cantonese dialect in Southeast Asia regard it as an auspicious plant in a variety of settings.
- Its branchlets with scale-like leaves are placed in the *hóng bāo* or red packets given to newly married couples in Cantonese wedding ceremonies. The branchlets serve as a blessing of good luck and happiness for the couple.
- It is commonly planted outdoors to bless the house and its occupants and to prevent misfortune in everyday life.
- Based on feng shui beliefs, the green and bushy form of the plant makes it symbolic of the wood element, so this plant is most auspicious for planting in the east, southeast or south zones of the garden, but inauspicious for the northeast and southwest, and neutral for the rest.
- Chinese *arbor-vitae* is grown as an ornamental (many cultivars, including colour and dwarf forms), for timber and used in traditional medicine.
- It is distinguished from the *Thuja* species in having fleshy female cone scales when the cone is immature (versus dry in *Thuja*) and wingless seeds (versus winged seeds).

Prickly, Spiny or Thorny Plants

Scientific Name	Common Names	Characteristics
Aegle marmelos	See page 4	Thorny
***Agave* species**	agave, century plant (English); *agave* (French); *Agave* (German); *lóng shé lán* [龙舌兰/龍舌蘭] (Mandarin); *agave* (Spanish)	Spiny
Ananas comosus	See page 8	Spiny
Cactaceae	Cactus family members, the cacti (English); *cactus* (French); *Kaktus* (German); *kaiktasa* (Hindi); *kaktus* (Malay); *xiān rén zhǎng kē* [仙人掌科] (Mandarin); *cactus* (Spanish)	Spiny
Ceiba pentandra	See page 26	Prickly
×*Citrofortunella microcarpa*	See page 38	Thorny
Citrus medica* var. *sarcodactylis	See page 40	Thorny
Citrus reticulata	See page 42	Thorny
***Crataegus* species**	See page 52	Thorny
***Euphorbia* species**	euphorbia (English); *euphorbes* (French); *Wolfsmilch* (German); *euphorbia* (Malay); *dà jǐ* [大戟] (Mandarin); *euphorbia* (Spanish)	Spiny
Fortunella japonica* or *Fortunella margarita	See page 76	Thorny
***Ilex* species**	See page 98	Spiny
Lycium barbatum* var. *barbatum	Barbary matrimony vine, Duke of Argyll's tea tree, goji berry, matrimony vine, wolfberry (English); *Borksdorn* (German); *níng xià gǒu qǐ* [宁夏枸杞/寧夏枸杞] (Mandarin)	Thorny
Lycium chinense* var. *chinense	Chinese boxthorn, Chinese matrimony vine, Chinese wolfberry, wolfberry (English); *lyciet de Chine* (French); *chinesischer Bocksdorn* (German); *kauki, koki* (Malay); *gǒu qǐ* [枸杞] (Mandarin)	Thorny
Onopordum acanthium	cotton thistle, heraldic thistle, Scotch thistle, Scottish thistle, woolly thistle (English); *chardon aux ânes* (French); *Eselsdistel* (German); *thistle* (Malay); *dà chì jì* [大翅蓟/大翅薊] (Mandarin); *cardo borriquero* (Spanish)	Spiny
Paliurus spina-christi	Christ's thorn, crown of thorns, Jerusalem thorn, garland thorn (English), *bīn zǎo* [滨枣/濱棗]	Spiny
Phoenix dactylifera	See page 132	Spiny
Punica granatum	See page 150	Thorny
***Rosa* species and hybrids**	See page 152	Prickly and spiny
Sarcopoterium spinosum	thorny burnet (English)	Thorny
Solanum mammosum	See page 166	Spiny
Ziziphus spina-christi	Christ's thorn jujube, crown of thorns, Jerusalem thorn, garland thorn, Syrian Christ's thorn (English)	Spiny

 Photos: (opposite) Hugh Tan Tiang Wah

Agave angustifolia 'Marginata' have sharp spines at the leaf tips

How to Grow: Please refer to specific species for plants found in this book or to other references. In general, the drawbacks of growing armed plants is that they are painful to handle, plant or prune, so special techniques such as holding a prickly stem with a rolled newspaper lasso or using leather gloves are necessary. If one were a practitioner of feng shui, one would have to carefully site them in the garden (preferably not grow them) and never bring them into the home.

Plant Description:

- All such plants are armed, i.e., they possess sharp structures growing from one or more plant parts.
- Prickles are hard, sharp outgrowths of the plant's stem surface but not developing from the leaf axil (the region demarcated by the upper part of the leaf where it is inserted and the twig or stem to which the leaf is attached), e.g., rose stems.
- Spines are hard, sharp outgrowths of the plant's leaves, including the stipules, leaf stalk, leaf sheath, leaf blade, leaflets and so on, e.g., holly leaves.
- Thorns are sharp outgrowths that develop from buds at the axil of the leaf, e.g., citruses.

Echinocactus grusonii has numerous spines

Euphorbia lactea spiny stem

Euphorbia milii cultivar flowers, leaves and spines

Sarcopoterium spinosum at Tel el Ful, Jerusalem, Israel

Sarcopoterium spinosum

Photos: (this page, clockwise from top) Hugh Tan Tiang Wah, Joseph Levi, Anna Sidiropoulou; (opposite top) Rafael Medina; (opposite bottom) Aila Ventura

Paliurus spina-christi winged fruits

Ziziphus spina-christi branch with flowers, northern Israel

Folklore, Beliefs, Uses and Other Interesting Facts

- The possession of prickles, spines or thorns by plants has often been associated with the ability to repel evil spirits or demons in many cultures.
- The Saxons of Transylvania hung branches of wild roses or other armed plants over their farm gates to frighten off witches who may sneak in by riding on the farmers' cows.
- The peach-wood guardian deities of Han Dynasty China—*shén shū* and *yù lù*, are thought to be derived from turmeric (*Curcuma longa*) and, possibly, the sow thistle (*Sonchus oleraceus*), respectively. The strong aroma of the former is meant to repel evil spirits but if they are stubborn and still try to enter the home, then the sow thistle is the last line of defence, being spiny, bitter and unpleasant.
- Armed plants are considered 'bad' feng shui plants because their sharp structures symbolise numerous 'poison arrows'—hostile, inauspicious energy which will harm the residents of the house. For this reason, they must not be placed inside the home. If one is a diehard fan of such plants, the compromise is to grow such plants as friendly sentinels by placing them outside in the garden, near the gate, or outside the wall or fence to guard the entrance to the home. Cacti or euphorbias (which resemble cacti), also suggest a desert landscape, which symbolises difficult living conditions, the opposite to what one hopes to achieve through the practice of feng shui. Particularly inauspicious plants are cacti or agaves, which resemble balls of numerous spikes.
- In the language of flowers, a thorny branch signifies severity and rigour. A rose without prickles represents a sincere friend but, on the other hand, black thorn (*Prunus spinosa*), which is spiny, represents difficulty or insouciance. Bramble (*Rubus* species) is prickly and represents remorse; a creeping cactus is spiny and represents horror; lantana (*Lantana camara*) is prickly and represents rigour; nettle (*Urtica* or other species with stinging hairs) represents cruelty, slander or sobriety; the thistle (*Carduus*, *Carlina*, *Cirsium* or *Onopordum* species) is spiny and represents austerity, criticism or misanthropy.
- Some species with armed branches have often been associated with the crown of thorns, the mock crown that Roman soldiers placed on Jesus prior to his Crucifixion at Calvary (Golgotha). These have included Christ's thorn (*Paliurus spina-christi*), crown of thorns (*Euphorbia milii*), hawthorn (*Crataegus* species), holly (*Ilex aquifolium*), Syrian Christ's thorn (*Ziziphus spina-christi*), thorny burnett (*Sarcopoterium spinosum*), and others.

Prunus mume • Prunus salicina

plum blossom • Japanese apricot • Chinese plum

Scientific Species Names:
Prunus mume **and** ***Prunus salicina*** cultivars
(Latin *prunus*, cherry or plum; *mume* is a variation of the Japanese *ume*, *Prunus mume*; Latin *salicina*, willow-like)

Common Species Name for *Prunus mume*:
Japanese apricot, mei, mume (English); *abricot du Japon, abricotier du Japon, abricotier japonais, prune du Japon, prune d'ume, prunier japonais, umé* (French); *Japanischer Aprikosenbaum, Japanische Aprikose, Mumebaum, Schneeaprikose* (German); *ume, ume no mi, miume* (Japanese); *méi* [梅], *wū méi* [乌梅/烏梅] (Mandarin); *albaricoquero japonés* (Spanish)

Common Species Name for *Prunus salicina*:
Chinese plum, Japanese plum (English); *prune du Japon, prunier du Japon, prunier Japonais* (French); *Japanischer Pflaumenbaum, Chinesischer Pflaumenbaum, Dreibluetige Pflaume* (German); *sumomo* (Japanese); *ijas jepang* (Malay); *rì běn lǐ* [日本李/歐洲李], *lǐ zǐ* [李子] (Mandarin); *cirolero japonés, ciruela japonesa, ciruelo japonés* (Spain)

Scientific Family Name:
Rosaceae

Common Family Name:
rose family

Natural Distribution:
Prunus mume: Southwestern Japan; *Prunus salicina*: China

USDA 6–9 (AUS 1–3)

How to Grow: Low maintenance, fast-growing, temperate tree; flowers best in full sun; deep, medium moisture, well-drained, slightly acidic loam soils are best; plant in a protected area for areas with strong winters as they are less cold-hardy; prune heavily to promote heavy flowering; propagation by grafting onto rootstocks, or by seed for breeding.

Drawbacks: Low late winter temperatures can damage flowers and, consequently, the fruit; plants do better in warmer temperate or Mediterranean zones; require chilling for flower development (averaging 550–800 hrs for *Prunus salicina*); windfall fruits can be messy; excessive rainfall during the fruiting season can result in fruit cracking or encourage diseases; diseases include bacterial canker, brown rot and *Verticillium* wilt; pests include aphids, borers, scale, spider mites and tent caterpillars.

Plant Description:

- A fast-growing, deciduous tree that grows to 6–9 m (19.7–29.5 ft) tall with a round crown or a large shrub (*Prunus mume*) or trees that grow to 9–12 m (29.5–39.4 ft) tall (*Prunus salicina*).
- The stalked, simple leaves have leaf blades that are broadly oval or egg-shaped (*Prunus mume*) or oblong-obovate to narrowly oval (*Prunus salicina*). The margins of the leaf blades are toothed (*Prunus mume*) or scalloped (*Prunus salicina*).
- The spicily fragrant flowers of *Prunus mume* have red sepals, pink to white petals, yellowish stamens and are arranged in clusters of 1–2. Flowers of *Prunus salicina* are white-petalled and develop in clusters of (usually) 3. Depending on the cultivated variety, the flower petals can be single (5 petals), semi-double (6–9 petals) or double (10 or more petals). Flowers bloom in late winter, before the leaves emerge.
- The *Prunus mume* fruit is a summer-ripening, round, fuzzy, green to yellow, apricot-like drupe, growing to 3.8 cm (1.5 in) across with a clinging stone. The *Prunus salicina* fruit is a round, egg-shaped or conical, usually yellow or red, sometimes green or purple drupe (like that of a plum) with a waxy bloom.

 Photos: (opposite top) Bradley Arnett; (opposite middle) Sing H. Lin; (opposite bottom) Hugh Tan Tiang Wah

Prunus mume cultivar blossoms

Prunus mume cultivar tree in Staten Island Botanic Garden, New York, USA

Choya umeshu (*Prunus mume* or plum liquor)

Folklore, Beliefs, Uses and Other Interesting Facts

- The plum blossom of traditional Chinese and Japanese culture refers to either of two species: *Prunus mume* or *Prunus salicina*. *Prunus mume* originates in southwestern Japan but has been cultivated for centuries in China and become deeply embedded in Chinese culture. *Prunus salicina*, cutivated for thousands of years in China, was introduced 200 to 400 years ago to Japan, from where it spread to the rest of the world and became known as the 'Japanese plum'.

- In traditional Chinese flower symbolism, the plum blossom epitomises the spirit of courage and resilience towards hardship. After a long hard winter, beautiful white-pink blossoms appear, even before the green leaves, and the tree suddenly comes back to life! The ability of the tree to survive and to bloom immediately after harsh winters, year after year, has made it the symbol of courage and chastity amongst the 'Four Gentlemen of Flowers'—*méi lán zhú jú* [梅兰竹菊/梅蘭竹菊]—and made it the representative of winter. The other three plants are the orchid (spring), bamboo (summer) and chrysanthemum (autumn).

- The plum blossom is also included in a triad of plants known as 'The Three Friends of the Cold Season'—*sùi hán sān yǒu* [岁寒三友/歲寒三友]. Together with the pine and bamboo, the plum blossom reiterates the never-say-die spirit even in the most adverse conditions. All three plants are able to conquer the obstacles and difficulties put forth to them during winter.

- To the Chinese, the plum blossom is associated with the Chinese New Year, sharing the equivalent status of the cherry blossom to the Japanese. People meet to view plum blossoms in favoured localities.

- In Southeast Asia, the Chinese purchase small potted cuttings of the flowering plum prior to Lunar New Year, although the climate is not suitable for its growth. Much effort is spent to keep the plant alive, and this includes placing it in a cool air-conditioned environment and watering it with ice-cold water. It is believed that luck and prosperity will be bestowed on the owner if the flowers of the plum do not die. The successful flowering of the plant in spring also affirms its ability to survive through the harsh winter by shedding its leaves, hence symbolizing the spirit of perseverance and the ability to negotiate tough situations, traits favored by the Chinese.

- The plum blossom and tree are symbols of good luck to the Chinese. Based on feng shui beliefs, it would be auspicious to plant a plum tree in the north (water) zone of the garden. This plant is also an important symbol of longevity.

- *Prunus mume* is the most widely cultivated and most often illustrated plum blossom of Japan. Its blooms, and those of the cherry blossom amongst others, are associated with spring. This plant is also used for bonsai and for its edible fruit.

- The winter-resilient *Prunus mume* (*Prunus mei*), the national flower of the Republic of China, has great symbolic significance for the Taiwanese. Its one-long-two-short stamen pattern signifies Dr. Sun Yat-sen's Three Principles of the People and its five petals, the government's branches. In 1982, the plum blossom was named as the city flower of Nanjing, People's Republic of China.

Prunus persica cultivars

peach • peach blossom

Scientific Species Name:
***Prunus persica* cultivars**
(Latin *prunus*, cherry or plum; Latin *persica*, of Persia (now Iran), referring to the mistaken idea that the peach originated there)

Common Species Name:
common peach, peach, peach blossom (English); *pêche*, *pêcher* (French); *Pfirsich* (German); *aru* (Hindi); *piichi*, *ke momo*, *momo* (Japanese); *persik* (Malay); *dà táo rén* [大桃仁], *máo táo* [毛桃], *táo* [桃], *táo rén* [桃仁], *táo zǐ* [桃子]; *melocotón*, *pérsico duraznero* (Spanish)

Scientific Family Name:
Rosaceae

Common Family Name:
rose family

Natural Distribution:
Only cultivated and not found wild, probably the first domesticated fruit tree in China.

USDA 5–8 (AUS 1–2)

How to Grow: A high maintenance, temperate fruit tree; benefitting from regular watering, fertilisation and pruning and periodic pesticide spraying; medium moist, well-drained, loamy to moderately sandy, slightly acidic soils; propagation by grafting onto rootstocks, and by seed only for breeding.

Drawbacks: Do not plant in the same soil which previously held other peach trees to prevent peach tree short life syndrome (possibly becoming predisposed to this by the ring nematode in the soil); prone to diseases including bacterial leaf spot, brown rot, canker, mildew, peach leaf curl, peach scab and root rot; pests include aphids, peach twig borer, plum curculio, oriental fruit moth, root nematodes, scale, spider mites; good horticultural and sanitation practices are necessary to prevent pest and disease spread; very low winter temperatures and cold spring frosts can significantly injure buds or flowers.

Plant Description:

- A deciduous tree that grows to 8 m (26.2 ft) tall. Dwarf cultivars grow to 2.4–3 m (7.9–9.8 ft) tall.
- The alternate, stalked, simple leaves have linear leaf blades with toothed margins, 5–15 cm (2.0–5.9 in) long.
- The spring-blooming, shortly stalked flowers are arranged in small groups of 1–2 on twigs. The petals can be white, pink to purple, and single (5 petals) or double (many petals) depending on the cultivated variety. There are numerous stamens surrounding a pistil consisting of a stigma, style and ovary. For flowers to develop, plants must be exposed to a period of low temperature (averaging 600–900 hours). Usually, flowers are self-pollinating.
- The fruit is a finely hairy, yellow blushed pink, orange or red drupe (like that of a plum) of up to 8 cm (3.1 in) across. Plants producing smooth-skinned, hairless fruits are known as nectarines (*Prunus persica* var. *nectarina*).

Peach blossoms

 Photos: (above) Dana Duncan Seil; (opposite top) Keith Weller, ARS, USDA (opposite middle and bottom) Hugh Tan Tiang Wah

Prunus persica 'Flameprince' fruiting branch

Potted peach plants sold during Chinese New Year

Peach fruit ceramic piece at the Kuan Imm Tong Hood Temple, Singapore

Folklore, Beliefs, Uses and Other Interesting Facts

- The peach blossom is an important part of Chinese folklore and tradition. This tree produces beautiful blossom-laden branches in northern China around February, usually the time of the Chinese New Year. It symbolises spring and is brought into the home as an auspicious decoration. Chinese New Year is an auspicious time for weddings and so, by association, the peach blossom also symbolises marriage and its flowers the face of a female beauty or a bride.

- The peach fruit is also a symbol of longevity. The specific name given to a peach which is expected to bless one with long life is *shòu táo* [寿桃]. The concept of the *shòu táo* is applied to common everyday items from peach-shaped pendants, curios, and landscape paintings with peach blossoms. It is even applied to food. Tasty longevity buns—*shòu bāo* [寿包/壽包] or simply 'longevity peach' (*shòu táo)* [寿桃/壽桃]—shaped and coloured like a peach fruit, are served to persons on their birthday to bless them with good health and a long life.

- The association of the peach with immortality stems from Chinese folk religion mythology. In the garden of the Western Royal Mother, *xī wáng mǔ* [西王母/西王姆], grows a peach tree whose fruits ripen every 3,000 years on her birthday. The Immortals of Kun Lun Mountains, *kūn lún shān* [昆仑山/崑崙山], gather to celebrate her birthday with a great feast, the Feast of Peaches—*pán táo huì* [蟠桃会/蟠桃會]. The fruits of the magical peach tree are served and they are believed to confer longevity to all who consume them. One of the Immortals, the God of Longevity, *shōu xīng* [寿星/壽星], is often depicted holding a peach or sitting on one.

- In China, the peach is also believed to protect against and repel evil. During the Chinese New Year, peach soup is drunk as protection against poisonous vapours and demons. A flowering peach branch is tied over gates (or water sprinkled with such a branch on the ground surrounding a house) to protect it from evil. Peach wood is thought to frighten away evil spirits, so children wear peach wood or peach pit amulets.

- The peach is also central to the Japanese folktale—*Momotarō*, the Peach Boy. An old, childless woman lived with her husband in the woods. One day, while her husband collected twigs in the woods, she went to the river to wash their clothes. A giant peach floated down the river and the old woman retrieved it and brought the peach home. She halved the peach and out came a boy. As the boy appeared from the peach, the old man named him *Momotarō*, *momo* for peach and *tarō*—a common name for a boy. The boy defeated devils which lived on Devil's Island with the help of his companions: the dog, the monkey and the pheasant.

- In the language of flowers, the peach blossom signifies constancy or says "I am your captive".

Prunus serrulata • *Prunus ×yedoensis*

cherry blossom • Tokyo cherry

Scientific Species or Hybrid Names:
Prunus serrulata and ***Prunus ×yedoensis*** cultivars
(Latin *prunus*, cherry or plum; Latin *serrulata*, small saw-toothed; × = multiplication sign, signifying the hybrid status; Latin *yedoensis*, of Yedo (Edo), now called Tokyo)

Common Species Name for *Prunus serrulata*
cherry blossom, oriental cherry, East Asian cherry (English); *cerisier à feuilles en dents de scie*, *cerisier du collines*, *cerisier du Japon* (French); *Gesägtblättrige kirsche*, *Grannenkirsche*, *japanische Blütenkirsche*, *japanische Kirsche*, *Nelkenkirsche* (German); *sakura*, *sato zakura*, *yama zakura* (Japanese); *bunga sakura*, *ceri* (Malay); *shān yīng huā* [山樱花/山櫻花] (Mandarin)

Common Species Name for *Prunus ×yedoensis*:
Potomac cherry, Tokyo cherry, Yoshino cherry (English); *cerisier Yoshino* (French); *Tokiokirsche*, *Yoshinokirsche* (German); *somei yoshino* (Japanese); *dōng jīng yīng huā* [东京樱花/東京櫻花], *rǎn jǐng jí yě yīng* [染井吉野樱/染井吉野櫻] (Mandarin)

Scientific Family Name:
Rosaceae

Common Family Name:
rose family

Natural Distribution:
Prunus serrulata: East Asia (China, Japan and Korea, extending to northern India in Sikkim and West Bengal); *Prunus ×yedoensis* is a possible hybrid between Oshima cherry or oshima zakura (*Prunus speciosa*) and higan cherry (*Prunus subhirtella*)

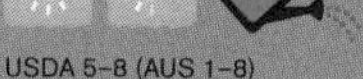

USDA 5–8 (AUS 1–8)

How to Grow: A high maintenance, spring flowering tree; does best in fertile, moist, well-drained loams; some cultivars are marginally winter hardy; propagation by stem cuttings or grafting.

Drawbacks: Prone to many diseases and pests; diseases include dieback, fire blight, leaf curl, leaf spot, powdery mildew and root rot; pests include aphids, borers, caterpillars, Japanese beetles, leafhoppers, scale, spider mites and tent caterpillars; prone to witch's broom.

Plant Description:

- A deciduous tree that can grow to 15–22.5 m (49.2–73.8 ft) tall in the wild (*Prunus serrulata*) or 9–13.5 m (29.5–44.3 ft) tall (*Prunus ×yedoensis*) when cultivated. Cultivars that are grown for their profuse spring blossoms are usually small trees.
- The alternate, stalked, simple leaves have egg- to lance-shaped (*Prunus serrulata*) or obovate (*Prunus ×yedoensis*) leaf blades with toothed margins, growing to 13 cm (5.1 in) long. Leaves turn from bronze to red-yellow (*Prunus serrulata*) or yellow (*Prunus ×yedoensis*) in autumn.
- The fragrant or non-fragrant flowers (*Prunus serrulata*) or fragrant flowers (*Prunus ×yedoensis*) are arranged in small clusters and are borne on twigs before leaves emerge in the spring. The petals can be white to pink and single (5 petals) or double (many petals; *Prunus serrulata* only) depending on the cultivated variety. There are numerous stamens surrounding a pistil, consisting of a stigma, style and ovary.
- The fruit is a small, magenta to black drupe (like that of a cherry or plum).

 Photo: (opposite) Scott Bauer, ARS, USDA

Cherry blossoms, Washington, D.C.

Folklore, Beliefs, Uses and Other Interesting Facts

- The most commonly cultivated flowering cherry in Japan is the *Prunus ×yedoensis*.
- The Japanese hold cherry blossoms in high spiritual regard. The exquisite blooms are short-lived and soon replaced by the bright green foliage in mid-spring. This is believed to reflect the importance of man leading a meaningful life even though it may be short.
- Much folklore surrounds the cherry blossoms. One particularly charming legend explains the magical, fleeting nature of flowering. It is said that as the frost of winter subsides, the maiden fairy *Ko-no-hana-sukuya-hime* wakes up the dormant cherry trees with her delicate breath and brings them back to life!
- The cherry blossom is Japan's unofficial national flower. People from all walks of life congregate at *hanami* (which literally translates to 'flower viewing') festivals to appreciate the blossoms amidst the company of friends and food. The practice of *hanami* is so popular in Japan that the national meteorological service provides yearly cherry blossom flowering forecasts!
- In 1912, the mayor of Tokyo, Yukio Ozaki, presented 1,800 *Prunus ×yedoensis* (Yoshino) and 1,220 *Prunus serrulata* cultivars to the then First Lady, Helen Herron Taft, wife of US President William Howard Taft, in appreciation for his support of the Japanese when he was the US Secretary of War during the Russo-Japanese War of 1904–1905. These trees were planted in Potomac and Rock Creek Parks and the White House grounds in Washington D.C. At the same time, Ozaki also presented 3,000 trees to New York City and some of these can still be found in Central Park, around Pilgrim Hill, and in the Conservatory Garden on 104th Street.

Prunus serrulata cultivar

 Photos: (above) Stephen Ausmus, ARS, USDA; (opposite) Elaine Chan

Cherry blossoms in Kamakura, Japan

- *Prunus serrulata* consist of several botanical varieties. The typical variety is *Prunus serrulata* var. *serrulata*. Japanese flowering cherry or *sato zakura* (*Prunus serrulata* var. *lannesiana*) is the one cultivated. Korean mountain cherry (*Prunus serrulata* var. *pubescens*) is found in China and Korea. Japanese mountain cherry, hill cherry or *yama zakura* (*Prunus serrulata* var. *spontanea*), the wild form of this species, is the cherry of Japanese painting and poetry. Extracts of this botanical variety have shown anticancer and antioxidative activities.
- Based on feng shui beliefs, the pink-flowered cherry trees are auspicious if planted near the house front, or the south, southwest and northeast zones of the garden; inauspicious for the west and northwest; and neutral for the rest. The white-flowered trees are most auspicious if planted in the west, northwest or north zones of the garden; inauspicious in the east and southeast; or in the front of the house, and neutral for the other zones.
- In the language of flowers, the cherry represents good education or says “Forget me not”.

Punica granatum pomegranate

Scientific Species Name:
Punica granatum
(Latin *Punica*, derived from *punicum malum*, the Carthaginian apple; Latin *granatum*, many-seeded)

Common Species Name:
pomegranate (English); *grenade* [fruit], *grenadier* [tree] (French); *Granatapfel* (German); *anaar*/*anar* (Hindi); *buah delima*, *delima* (Malay); *shí líu* [石榴] (Mandarin); *granada*, *granado*, *mangrano* (Spanish)

Scientific Family Name:
Lythraceae (formerly in Punicaceae)

Common Family Name:
loosestrife family (formerly in pomegranate family)

Natural Distribution:
Turkey to the Himalayas in Pakistan and Northern India. Cultivated in the Mediterranean region since ancient times by the Greeks and Romans, and now in the subtropics and tropics.

USDA ≥8 (AUS ≥2)

How to Grow: The pomegranate plant grows best with hot, dry summers and cool winters; it thrives best in sandy, clayey, acidic and even alkaline soils; propagation by stem cuttings, air-layerings or seed (but offspring may not resemble the parent plant).

Drawbacks: It has a nice conical shape when young, but as it grows, it will require pruning to avoid a weeping shape; relatively free from pests and diseases; minor problems include fruit and leaf spot and leaf damage by fungi, white flies, thrips, mealybugs and scale insects.

Plant Description:
- A branching shrub that grows to about 2 m (6.6 ft) tall.
- The opposite or clustered, shortly stalked leaves are 1–9 cm (0.4–3.5 in) long, and leaf blades are oblong-lanceolate, glossy and light-green.
- Flowers are bright red, up to 3 cm (1.2 in) in diameter, growing clusters of 1–5. The calyx is leathery and bell-shaped, and the 3–7 petals are crinkled, white, red or variegated. Stamens are fine and numerous with the style growing beyond them.
- The fruit, up to 12 cm (4.7 in) across, is a somewhat round berry, brownish-red to purplish-red when ripe.
- There are numerous seeds inside the hard, leathery fruit wall. These seeds are surrounded by an edible, tart-flavoured, juicy, waxy-looking, pink or pale yellow, translucent aril (external covering of the seed).

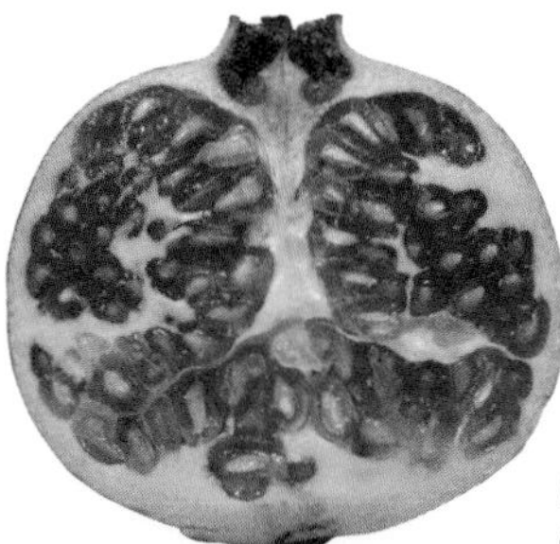

Punica granatum whole fruit and fruit in section

 Photos: (above and opposite) Hugh Tan Tiang Wah

Potted *Punica granatum* bush

Flower and buds on a leafy twig

Young fruit on branch

Folklore, Beliefs, Uses and Other Interesting Facts

- The pomegranate is associated with the ancient Greek legend of Persephone, who was abducted by Hades, the dark god of the underworld, to be his queen. Her mother, Demeter, Goddess of the Harvest, grieved for her daughter, so everything ceased to grow. Zeus compelled Hades to return Persephone to Earth. However, Hades tricked Persephone to eat some pomegranate seeds. (It was a rule that anyone who ate or drank in the underworld was condemned to spend eternity there.) In one version of the myth, Persephone consumed four seeds of the pomegranate. She was thus condemned to spend one month in the underworld with Hades for every seed she ate. Every year, when Persephone is confined in the underworld for four months, Demeter mourns and the earth becomes less fertile, explaining the barren winter season.
- This association of the pomegranate with the cycle of birth and death made it the Christian symbol of the resurrection. The plant is also a Christian symbol of God's abundant provision.
- In the Torah, the first commandment is to be fruitful and multiply, so, to the Jews, the pomegranate is a symbol of fertility.
- In Arabian folklore and poetry, the pomegranate represents the female breast, which is also associated with nurture and fertility.
- The pomegranate is one of the sacred fruits of Buddha. In Buddhist legend, Buddha gave the children-devouring demoness Hariti a pomegranate and, after eating it, she was reformed. The sanctity of the pomegranate, exemplified by its ability to 'cure' an evil demoness, makes it a popular auspicious plant especially among the Buddhists and Taoists. It is cultivated in temples and shrines. The fruits are harvested to be used as offerings on the altar and the leaves are packed together with other flowers to be sold (for a nominal fee) or given as bath flowers.
- To the Chinese, the pomegranate's numerous seeds represent many sons earning fame and glory, so it is a symbol of fertility and prosperity. The leaves of the pomegranate are soaked in water preceding a bath, so that the leaf-soaked water can 'cleanse' the body of bad luck.
- Based on feng shui beliefs, the pomegranate, with its bright red flowers, is auspicious if planted in the south, southwest or northeast sectors of the garden; inauspicious if planted in the west or northwest; and neutral for the rest. Some varieties of the plant are thorny, so such plants should not be grown too close to the house, and certainly not inside, as the thorns represent poison arrows sending out inauspicious energy in all directions. However, thorny plants can also act as 'sentinels', so it is commonly cultivated by the Chinese as a pot or garden shrub to counter evil forces. Non-thorny plants are also auspicious if placed at the front of the house.
- The pomegranate fruit is eaten raw; its juice is used in cooking; a concentrated pomegranate syrup (grenadine) is used to make drinks.

Rosa species & hybrids

garden rose

Scientific Species or Hybrid Names:
***Rosa* species** and **hybrids**
(Latin *rosa*, rose; there are about 200 *Rosa* species worldwide)

Common Species or Hybrid Names:
rose, garden rose (English); *rose* (French); *Rose* (German); *gul* (Hindi); *ros* (Malay); *méi* [玫], *méi guì* [玫瑰], *qiáng* [蔷], *qiáng wēi* [蔷薇/薔薇] (Mandarin); *rosa* (Spanish)

Scientific Family Name:
Rosaceae

Common Family Name:
rose family

Natural Distribution:
Northern temperate countries to tropical mountains

USDA ≥8 (AUS ≥2)

How to Grow: *Rosa* cultivated varieties of species and hybrids are numerous, and they grow over a wide range of climates from cool temperate to subtropical or tropical montane, or even tropical, so it is best to determine the individual requirements of specific kinds from the relevant references; fast growing; tolerates a wide range of soil types and pH levels but must be well-drained; medium tolerance for drought and salt; propagation by stem cuttings, grafting, tissue culture or seed.

Drawbacks: Prickly plants so unpleasant to prune or work around; black leaf spot disease; insect pests include red spider mites and chewing insects.

Plant Description:

- The garden roses commonly found in gardens and florists
- are cultivated varieties of species or hybrids.
- They are erect, climbing or creeping shrubs with solitary
- or clustered flowers at the ends of branches.
- Prickles are usually present on the stem and branches. The 'thorns' that roses are said by laypersons to possess are technically known as prickles, since they are found all over the stem and not found at the axil of the leaf, which is where thorns develop.
- The alternate, spiny, stipulate leaves are either imparipinnate or simple, and the blade of the leaf and leaflets possess toothed margins.
- Single flowers or inflorescences develop at the branch or stem tips. Wild type roses have 5 (rarely 4) sepals, 5 (rarely 4) petals, numerous stamens and numerous (rarely few) pistils, but 'double' flowers have more petals that are transformed from the stamens.
- The fruit, or rose hip, develops from the cup-like hypanthium that surrounds the base of the flower. It is a source of vitamin C. Inside this structure are the true fruits which develop from individual pistils, each containing 1 seed.

Rosa cultivar flower

 Photos: (above and opposite) Hugh Tan Tiang Wah

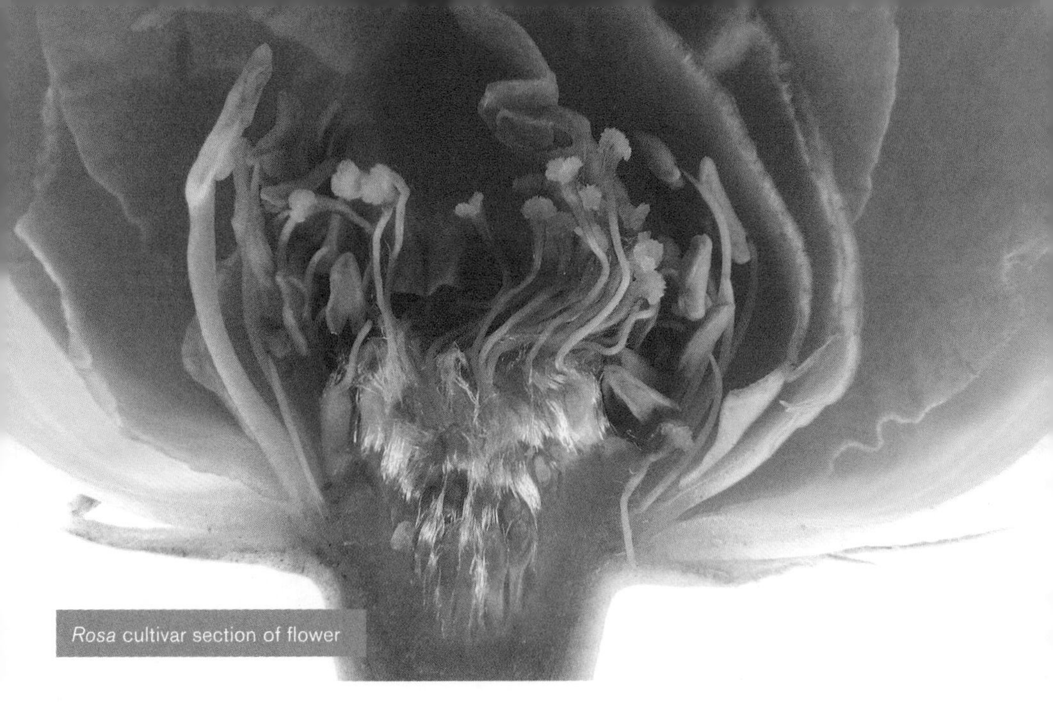
Rosa cultivar section of flower

Rosa cultivar flower

Folklore, Beliefs, Uses and Other Interesting Facts

- The red rose is the symbol of England or love (especially on St Valentine's Day or Valentine's Day) and worn occasionally on St George's Day (23 April).
- The rose was sacred to Venus in earlier times but later became the flower of the Virgin Mary, signifying virginity. Rose petals were used to line paths taken by processions honouring the Virgin Mary.
- Based on feng shui beliefs, the rose is neither a good nor bad plant. The beauty and bright colours of the flowers bring precious energy, but the prickles emit *shā qì* or hostile energy. Because of the latter, rose plants are better placed away from the house in flower beds or pergolas and close to the gate where they can act as guards, since the Chinese believe that they ward off evil spirits because they possess prickles, which the spirits fear.
- In the UK, it is considered unlucky to scatter the petals of a rose flower that was held in the hand or that was worn. But it is acceptable to scatter rose flowers on a grave or on the ground in a ceremony. A rose, or any other flower, that blooms out of season is considered a bad omen. In Worcestershire, it is believed that if green leaves develop between the petals of a red rose flower, this is an omen that one of the rose plant's owner's family members will die.

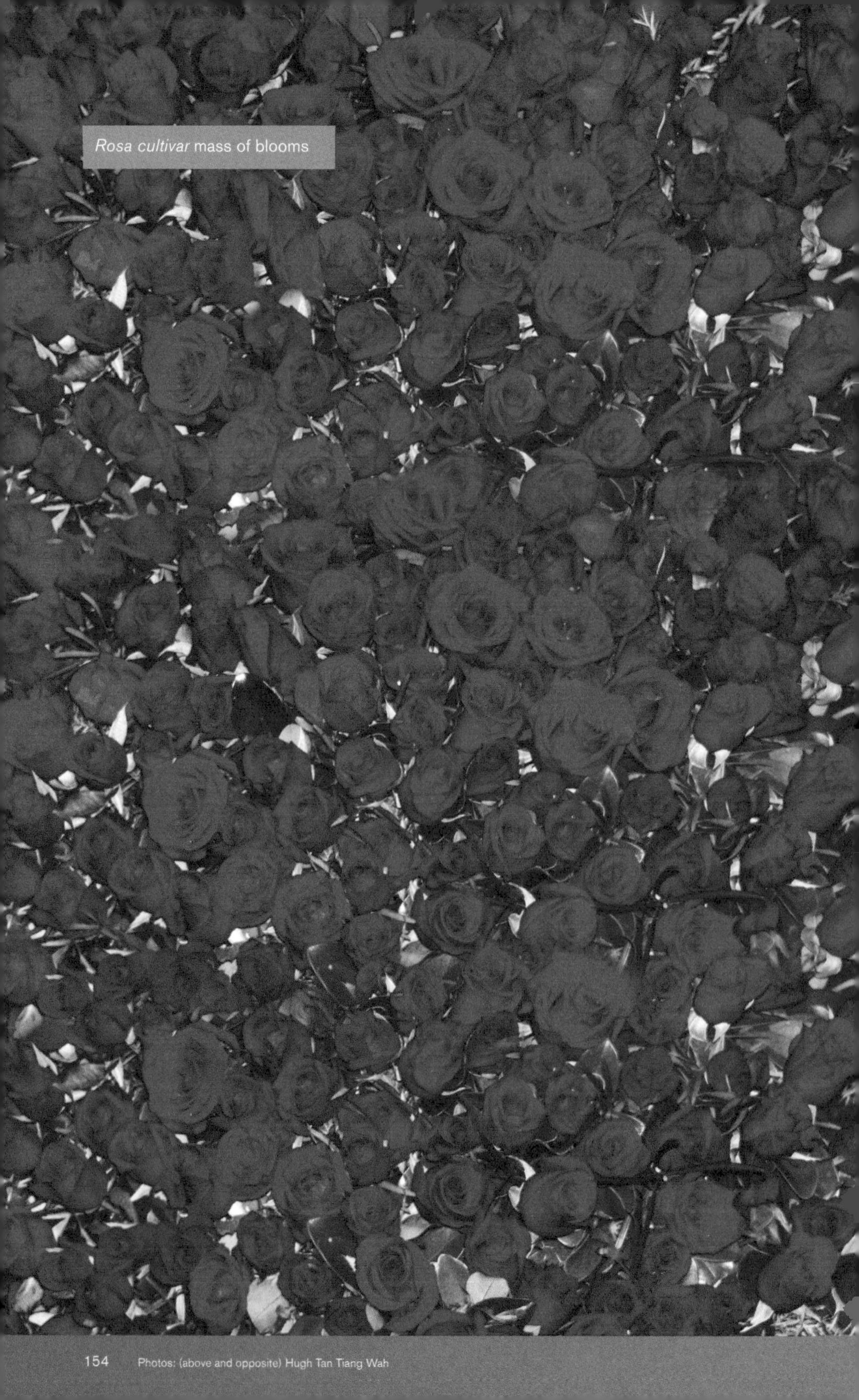
Rosa cultivar mass of blooms

 Photos: (above and opposite) Hugh Tan Tiang Wah

Rosa cultivar upper side of leaf

Lower side of leaf

Prickles on stem

- In the language of flowers, the rose stands for beauty, love or ephemeral beauty. A bicolour rose flower symbolises scandal or scholarship. A bridal rose signifies happy love. A burgundy (dark purplish red or blackish red) rose means unconscious beauty. A cabbage rose (*Rosa* ×*centifolia*) signifies the ambassador of love. A daily rose means to say "That smile I would aspire to". A damask rose (*Rosa* ×*damascena*) means freshness. A deep red rose signifies bashful shame. A monthly rose symbolises beauty always new. A red and white rose stands for sufferings of the heart or fires of the heart. A striped rose symbolises summer. A rose without prickles signifies a sincere friend. A white rose signifies silence or innocence, a white rose being the flower of Harpocrates, the god of silence. A dried white rose says "Better to die than to lose one's innocence" or transient impressions. A wild (wild type) rose represents simplicity. A withered rose means fleeting beauty. A yellow rose signifies infidelity. A crown of roses symbolises reward of virtue. A rose leaf says "I will not beg".
- The rose is the national flower of Bulgaria (red rose), Czech Republic (traditional), Ecuador, England (red tudor rose), Iran (red rose), Iraq, Italy (traditional), Luxembourg, Rumania (dog rose), Turkey (traditional) and the USA. Various kinds of roses are the state flowers of Georgia (*Rosa laevigata*), Iowa (*Rosa arkansana*), New York (*Rosa* species), North Dakota (*Rosa arkansana*) and Washington, D.C. (*Rosa* 'American Beauty') of the USA. The wild rose (*Rosa acicularis*) is the provincial flower of Alberta, Canada. City of Roses is a common title attributed to cities including Aberdeen, Scotland (UK), Chico and Pasadena, California (USA), Portland, Oregon (USA), Windsor, Ontario (Canada), Fukuyama (Japan) and many more.
- The rose is the birth month flower for June.
- Roses are commercially grown as ornamental plants or for their cut flowers, and thousands of cultivated varieties have been bred.
- Rose flowers are an important source of essential oils for perfumery and medicinal usage.

Rosmarinus officinalis

rosemary • friendship bush

Scientific Species Name:
Rosmarinus officinalis
(Latin *ros*, dew; Latin *marinus*, maritime; referring to the natural habitat of rosemary on the sea cliffs of southern Europe; Latin *officinalis*, sold in shops, this applying to plants with real or imagined medicinal properties)

Common Species Name:
compass plant, friendship bush, plant of friendship, rosemary (English); *incensier*, *romarin*, *romarin commun* (French); *Rosmarin* (German); *rusmari* (Hindi); *mí dié xiāng* [迷迭香] (Mandarin); *romero*, *romero comun*, *rosmario* (Spanish)

Scientific Family Name:
Labiatae or **Lamiaceae**

Common Family Name:
mint family

Natural Distribution:
Mediterranean region; widely introduced elsewhere

USDA 8–10 (AUS 2–4)

How to Grow: A Mediterranean plant; mean annual temperatures from 6–12°C (42.8–53.6°F) to 18–24°C (64.4–75.2°F) and mean annual rainfall of 1,000–2,000 mm (39.4–78.7 in) and 500–2,000 mm (19.7–78.7 in); soil pH of 4.5–8.3, but preferably 6–7.5; preferring calcareous (chalky) well-drained soils; moderate watering needed to prevent root rot; not frost tolerant unless specific cultivars; drought-resistant once established; propagation easiest by stem cuttings, plant division or air layering, difficult by seed because seed development occurs only under good growing conditions and seeds are usually only 10–20 per cent viable.

Drawbacks: Insect pests include the bugs *Phytocoris rosmarini* and *Orthothtylus ribesi*; fungal disease include *Sclerotinia sclerotiorum*; nematode root knot disease caused by *Meloidogyne incognita*.

Plant Description:

- An evergreen, perennial, erect to creeping shrub that grows to 2 m (6.6 ft) tall.
- The young branches are square in cross section and have fine grey hairs.
- The stalkless or shortly stalked leaves grow towards the upper portions of the branches, with the leathery leaf blade 1–5 cm (0.4–2.0 in) long and 1–2 mm (0.04–0.08 in) wide. The leaves are shiny dark green above and densely white felted below, the margins rolled under. Leaves are fragrant when bruised.
- The flowers occur in 5–10-flowered shoots at the tips of short branches. The stalked flowers have usually pale blue or blue, rarely white, petals which are joined into a 2-lipped structure, up to 7 mm (0.3 in) long.
- Each flower produces 4 nutlets (tiny nuts), each about 2 mm (0.08 in) long.

 Photos: (opposite top) Kate Kruger; (opposite middle and bottom) Hugh Tan Tiang Wah

Rosmarinus officinalis flowers

Leafy shoots

Potted plants sold in nurseries

Folklore, Beliefs, Uses and Other Interesting Facts

- In Europe, since the last millennium, rosemary has signified happiness, faithfulness and love, and has been used in weddings and funerals.
- Roy Vickery compiled reports from England that rosemary was the friendship bush or plant of friendship, and one would not be short of friends if it was planted in the garden. Rosemary was also reputed to offer protection from witches if planted near one's house. Those who wore it were protected from assault and physical injuries, evil spirits, fairies, lightning and witches.
- In the British Isles, traditional folklore has it that the originally blue flowers turned white when the Virgin Mary spread the Christ Child's linen to dry on a rosemary bush during her escape into Egypt. Another traditional tale has it that the plant grows for 33 years until it reaches Jesus Christ's height when He was crucified, after which it only grows broader.
- In ancient Greece, rosemary was believed to be a gift from the goddess Aphrodite to humans and was used as an incense substitute for altar offerings. Rosemary garlands were also placed on the statues of the gods.
- In the language of flowers, rosemary signifies good faith, fidelity or remembrance, and says "Your presence refreshes me". Because it signifies remembrance, it is used for both weddings and funerals. Until the 20th century, mourners in the British Isles dropped sprigs of the plant into the grave as the coffin was lowered to signify that the departed would not be quickly forgotten.
- Rosemary is a commonly cultivated ornamental plant and herb whose leaves and stems are used, fresh or dried, to flavour biscuits, bread, cream soups, herbal butters, jams, roast meats, sauces, sausages, stews and vegetables. The volatile oil (including verbenone, the character-impact compound), distilled from its flowering shoots and leaves, is used in perfumes and for scenting detergents, soaps, etc. and, together with the oils of bergamot (*Citrus aurantium ssp. bergamia*) and neroli (*Citrus aurantium*), is the main component of Eau-de-Cologne. This plant is also used in traditional medicine for asthma, baldness, bruises, colds, headaches and sore throats. Rosemary plants also provide nectar for honeybees and is an excellent honey plant.

Saccarum officinarum sugarcane

Scientific Species Name:
Saccarum officinarum
(Greek *sakcharon*, the sweet sap of the sugar cane, possibly derived from the Malay *singkara*; Latin *officinarum*, of shops, generally those of druggists)

Common Species Name:
noble cane, noble sugar cane, sugarcane (English); *canne à sucre* (French); *Zuckerrhor* (German); *ganna, gannaa, iikh, sakhara, ukh* (Hindi); *tebu* (Malay); *gān zhè* [甘蔗] (Mandarin); *caña de azúcar, caña de castilla, caña común, caña dulce, cañaduz, caña melar, cañamiel, caña sacarina* (Spanish)

Scientific Family Name:
Poaceae or **Gramineae**

Common Family Name:
grass family

Natural Distribution:
New Guinea, where it has been known to occur since 6,000 BC, and spread to the Malay Archipelago, India, China, Hawaii, Mediterranean, Caribbean and the Americas. Now cultivated in about 70 (mainly tropical) countries.

USDA ≥9 (AUS ≥3)

How to Grow: Tropical and subtropical plant; optimum temperatures for vegetative growth 30–33°C (86–91.4°F); rainfall of 1,800–2,500 mm (70.9–98.4 in) per year is best; uniform and high rainfall is best for vegetative growth; tolerates most soil types but prefers deep, fertile, organic-rich, friable, well-drained soils of pH 5–8; requires large amounts of nitrogen, and also significant quantities of potassium, phosphorus, calcium and silica, as it is grown for its vegetative parts; propagate by mature stem cuttings, and seed only for breeding new cultivated varieties.

Drawbacks: A large, untidy plant; silica cells in the leaves can scratch or cut, so should be handled with gloves; diseases include mosaic virus disease, ratoon stunting disease, yellow spot and rust; pests include stem borer, top borer, woolly aphid and rat.

Plant Description:

- A perennial grass that grows to 6 m (19.7 ft) tall.
- The robust, jointed stem (culm) is 2–5 cm (0.8–2.0 in) wide, and contains sweet sap, the source of cane sugar.
- The alternate, two-rowed leaves, each with a sheath which encircles the stem and a thin, narrow leaf blade 70–200 cm (27.6–78.7 in) long by 3–7 cm (1.2–2.8 in) wide, centrally thick and paper thin at the edges, rolling up to reduce surface area for transpiration when the plant is under water stress.
- The tiny flowers are borne in a branched, 25–50 cm (9.8–19.7 in) long inflorescence (tassel) at the stem tip.
- The fruit is a 1 mm (0.04 in) long grain (caryopsis).

 Photos: (opposite top) Scott Bauer, ARS-USDA; (opposite middle) Hugh Tan Tiang Wah; (opposite bottom) Derek Chew

Tassled sugar cane plants near Canal Point, Florida, USA

Saccharum officialis tops of leafy stems

Leafy stems (right) and a roast pig as a Taoist prayer offering

Folklore, Beliefs, Uses and Other Interesting Facts

- Sugarcane is considered sacred to the Taoist Jade Emperor of Heaven, 天公, *ti gong* (Hokkien) or *tiān gōng* (Mandarin), and its cultivation is believed to be auspicious. Hence, it is often planted in the gardens of the Chinese.
- A popular Chinese folktale describes people surviving from the Japanese air-raids during the Second World War by escaping from their homes and hiding in dense sugarcane plantations, further venerating the sugarcane as a lucky plant.
- When sugarcane is not cultivated at home, the culms are obtained from markets and placed at the altar by the Chinese for worship.
- The bow of Kamadeva, the Hindu God of Love, is occasionally represented by a sugarcane stem with a string of bees and his five arrows tipped by flowers.
- In the sugarcane plantations of the West Indies, there are many superstitions pertaining to this plant. One belief is that if the plant produces flowers and seeds again at the end of the growing season, it means that bad luck will befall the plantation owner or his family.
- According to Tahitian folklore, the sugarcane was made from the human backbone, which accounts for the joints in the stem.
- A Solomon Island myth has it that a man and a woman, humankind's parents, emerged from two different knots growing on a sugarcane stem.
- Sugarcane is cultivated for the sucrose in its stems, used in various food products, beverages or for making rum. The fibrous residue (bagasse) left over after juice extraction is used as fuel, for fibre and particle boards, plastics, paper and furfural. The molasses remaining after removal of the sugar crystals is used for animal feed, fertilizer, production of yeast, carbon dioxide, animal acids for animal feeds, drinking and industrial ethanol (the latter being used recently as a petrol substitute for cars).

Salix caprea pussy willow

Scientific Species Name:
Salix caprea
(Latin *salix*, willow; Latin *caprea*, pertaining to goats)

Common Species Name:
goat's willow, great sallow, pussy willow (English); *saule blanc*, *saule marsault* (French); *Salweide*, *Weidenkätzchen* (German); *dedalu kambing* (Malay); *huáng huā liŭ* [黄花柳] (Mandarin); *sauce blanco* (Spanish)

Scientific Family Name:
Salicaceae

Common Family Name:
willow family

Natural Distribution:
Europe to Northeast Asia

USDA ≥2 (AUS ≥1)

How to Grow: Check out the individual needs for specific cultivars from other references as this is a general species account; a fast-growing, hardy tree; most soil types; waterlogging, drought, salt aerosol, soil salt tolerant; propagate by stem cuttings or seed.

Drawbacks: Drooping branches of cultivars with a weeping form can obstruct pedestrian or vehicular traffic, so they should be pruned regularly along paths or roads, or not planted at the sides; weak wood so prone to breakage; roots can lift pavements or obstruct mowing, damage sewer and water pipes or septic tanks; insect pests include aphids, borers, caterpillars (*Salix*, the favourite of the gypsy moth) and scale insects; diseases include black canker, crown gall, leaf spot, root rot, rust, tar spot and willow scab; short-lived, usually to 30 years, but can be pruned to develop central trunk to increase the lifespan.

Plant Description:

- A deciduous shrub or small tree.
- Some cultivated varieties (e.g., Pendula, Weeping Sally) have drooping branches.
- The alternate, stalked leaves have leaf blades that are ovate-oblong, broadly ovate to obovate-oblong, 5–7 cm (2.0–2.8 in) long and 2.5–4 cm (1.0–1.6 in) wide, hairy below, smooth and hairless above, with the margins irregularly notched, toothed to slightly toothed.
- Plants are male or female. Flowers are borne in flowering shoots called catkins. Male catkins are 1.5–2.5 cm (0.6–1 in) long and 1.5 cm (0.6 in) across, consisting of an axis on which black bracts (modified leaves) are found at the tip and lighter below, the male flowers inserted in the axils, each consisting of 1 gland and 2 stamens with yellow anthers. Female catkins are 2–6 cm (0.8–2.4 in) long by 0.8–1.8 cm (0.3–0.7 in) across, with bracts like those of the male catkins, and the female flowers each consist of 1 gland and 1 pistil.
- The fruit is a capsule that is up to 9 mm (0.4 in) long and which splits into 2 parts to release the fine hair-covered seeds.

Photos: (opposite top) Peter Birch; (opposite middle) Giam Xingli; (opposite bottom) Hugh Tan Tiang Wah

Salix caprea female flowers (left) and developing fruit in catkins

Male flowering branch and sprouting leaves

Cut flowers for sale during Chinese New Year

Folklore, Beliefs, Uses and Other Interesting Facts

- The budding shoot of the pussy willow is a perennial auspicious favourite among the Chinese during their New Year celebrations who believe that in order to establish the start of the New Year, the pussy willow must grow new leaves to signify the breaking of winter and the coming of spring. As such, it is often exported before new leaves emerge. The Chinese living in warmer climates, such as Southeast Asia, will try their utmost to create the conditions conducive for its leaves to sprout; they do this by placing the branches in cold water and in an air-conditioned environment. If young green leaves sprout, spring is established and luck is guaranteed for the rest of the year.
- In the rural parts of the United Kingdom, the pussy willow branch is a substitute palm leaf for Palm Sunday. In Russia, the week before Palm Sunday is known as Willow Week.
- Pussy willow is grown for its budding shoots for cut flowers and as an ornamental plant. It is grown for erosion control, shade, tannin and wood. Medical folklore includes its use as an astringent or tonic and as a treatment for fever and piles.
- Based on feng shui beliefs, the 'weeping' form cultivated varieties are not auspicious to cultivate in a home garden because the drooping branches signify sorrow and the dense crown blocks the flow of *qì* (energy or life force) allowing stale *qì* to accumulate. A willow can also grow large and its crown may engulf the house, symbolising sadness overtaking all.
- The American pussy willow (*Salix discolor*) is similar in appearance and use.

Salix species & hybrids

willow • weeping willow

Scientific Species or Hybrid Names:
***Salix* species** and **hybrids**
(Latin *salix*, willow; there are about 520 *Salix* species and numerous hybrids. The weeping willows include *Salix babylonica* and the hybrid *Salix* ×*sepulcralis*, thought to have the former as a parent. (Latin *babylonica*, Babylonian; ×, the multiplication sign denotes the hybrid status; Latin *sepulcralis*, growing near burial sites)

Common Species Name:
willow (English); *saule* (French); *Weide* (German); *dedalu, willow* (Malay); *liǔ* [柳] (Mandarin); *sauce* (Spanish)

Common Species Name:
weeping willow (English); *saule pleureur* (French); *Trauerweide* (German); *chúi liǔ* [垂柳] (Mandarin); *sauce llorón* (Spanish)

Scientific Family Name:
Salicaceae

Common Family Name:
willow family

Natural Distribution:
Most species originate in the cold and temperate northern hemisphere with a few in the southern hemisphere.

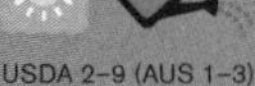

USDA 2–9 (AUS 1–3)

How to Grow: Check out the specific needs of individual species from other references as this provides only a general account for the willows; a fast-growing, hardy tree; most soil types; waterlogging, drought, salt aerosol, soil salt tolerant; propagate by stem cuttings or seed.

Drawbacks: Drooping branches of species and hybrids of the weeping willow can obstruct pedestrian or vehicular traffic, so should be pruned regularly along paths or roads or not planted at the sides; weak wood so prone to breakage; roots can lift pavements or obstruct mowing, damage sewer and water pipes or septic tanks; insect pests include aphids, borers, caterpillars (*Salix*, the favourite of the gypsy moth) and scale insects; diseases include black canker, crown gall, leaf spot, root rot, rust, tar spot and willow scab; short-lived, usually to 30 years, but can be pruned to develop the central trunk to increase the lifespan.

Plant Description:

- Usually deciduous, rarely evergreen, shrubs or trees.
- Some species and hybrids have characteristically long, slender, gracefully drooping branches and twigs (giving these species or hybrids the 'weeping tree' appearance).
- The usually alternate, stalked leaves have leaf blades that are usually long and narrow, developing usually on the vigorous small branches or twigs.
- Plants are unisexual with male and female flowers found on different plants in flowering shoots called catkins. Flowers lack sepals and petals, but each has 1 or 2 glands. Female catkins consist of bracts subtending pistils, each of which is from 1 female flower. Male catkins are short, consisting of bracts and 2 to many stamens per male flower.
- The fruit is a capsule which splits into 2 when ripe to release the silky plumed seeds that are wind-dispersed.

 Photos: (opposite) Hugh Tan Tiang Wah

Salix babylonica tree

Leafy branches

Folklore, Beliefs, Uses and Other Interesting Facts

- Taoists and practitioners of Chinese folk religion in Asia believe that the weeping willow has the power to ward off evil spirits and bad luck.
- A twig with leaves is often placed in homes, over the front doorway or elsewhere, to protect the family against evil spirits.
- Priests of Chinese folk religion often include twigs of the weeping willow in exorcism rituals. They use twigs with intact leaves to sprinkle holy water on the victim of spirit possession.
- Willow branches are used as a palm leaf substitute for Palm Sunday in many parts of Europe.
- Based on feng shui beliefs, any plant with a 'weeping' form, such as the weeping willow, is not auspicious to cultivate in a home garden because its drooping branches signify sorrow and the dense crown blocks the flow of *qì* (energy or life force), allowing stale *qì* to accumulate. A willow can also grow large and its crown may engulf the house, symbolising sadness overtaking all.
- The Babylon weeping willow (*Salix babylonica*) has a misnomer in its specific epithet, *babylonica*. Carl von Linné (Carolus Linnaeus), the father of modern taxonomy who named the plant, mistook its origin as Babylonia (the ancient country in the valley of the lower Euphrates and Tigris Rivers, now in modern-day Iraq) in Southwest Asia. The actual origin is probably China. The 'willows' mentioned in the Bible, in the waters of Babylon, are now thought to be the Euphrates poplar (*Populus euphratica*), a relative: "By the rivers of Babylon, there we sat down, yea, we wept, when we remembered Zion. We hanged our harps upon the willows in the midst thereof..." (Psalm 137:1–2, King James Version). Thus probably arose the association of weeping with the drooping branches and how the willow became a traditional emblem of grief in the West.
- In the British Isles, in the 16th and 17th centuries, the willow became associated with forsaken love. That a forsaken lover should wear a willow wreath or hat was a belief that lasted for centuries. It is thought by some that willow catkins are unlucky to bring into the home. However, if they are brought into the home on May Day, they protect the inhabitants from the 'Evil Eye' particularly if the catkins were provided by a friend. To beat a child or an animal with a willow cane will stunt his/her growth or cause internal pains, respectively, because willow wood decays quickly.
- In the language of flowers, the weeping willow signifies bitter sorrow, forsaken love, melancholy or mourning.

Saraca asoca asoka tree

Scientific Species Name:
Saraca asoca
(= *Saraca indica*) (*Saraca* is apparently a corrupted version of the East Indian *Asoka*, which refers to this tree; East Indian *asoca*, name of this tree; Latin *indica*, of India)

Common Species Name:
ashok tree, ashoka, ashoka tree, asok, asoka, asoka tree, sorrowless tree (English); *arbre ashoka* (French); *Ashokbaum* (German); *ashok*, *asok*, *sita ashok* (Hindi); *gapis*, *talan*, *tengalan*, *tanglin* (Malay); *wú yōu huā* [无优花/無優花] (Mandarin)

Scientific Family Name:
Fabaceae or **Leguminosae**
(formerly in the *Caesalpiniaceae*)

Common Family Name:
bean family

Natural Distribution:
Eastern Pakistan, Southern India, Sri Lanka, Assam, Upper Myanmar and Peninsular Malaysia

USDA >11 (AUS >5)

How to Grow: Tropical or subtropical evergreen tree; rainfed but better irrigated; well manured soil, with compost or vermicompost; propagation by seed and grafting.

Drawbacks: May become untidy in appearance; flowers may leave a mess; strong shade thrown by the crown may kill grass beneath.

Plant Description:
- A small, erect, evergreen tree.
- The alternate, nearly stalkless, pinnate leaves each bears 4–6 pairs of leathery, oblong or lance-shaped leaflets that are about 10–20 cm (3.9–7.9 in) long. Young leaves are limp and hang downwards.
- The flowers are arranged in a nearly round, branched cluster. The flowers are petalless, with 4 oval, petal-like sepals at the top of a tube, changing from yellow to orange then scarlet. The 4–8 slender stamens project beyond the sepals. The pistil consists of a stalk bearing the ovary, a thin style that curls into a ring and is tipped by the stigma.
- Fruits are flat, leathery legumes (pods) 15–25 cm (5.9–9.8 in) long. When ripe, they each split to release 4–8 seeds.

Saraca asoca young tree

Flowers

Fruit and young leaves

Folklore, Beliefs, Uses and Other Interesting Facts

- The asoka is sacred to Buddhists and Hindus. It is especially sacred to Buddhists because, according to one account, Gautama Buddha was born under an asoka tree.
- According to Hindu belief, Siva (the God of Destruction of the Hindu triad of gods) blessed the asoka to be immortal because his wife Parvati worshipped him with its flowers, so thus it became evergreen. Hindus worship the asoka on the 13th day of the month of Chaitra, and its flowers are used for religious offerings. In Bengal, women eat its blossoms on Asoka Shasthi day. Generally, Hindu women believe that their children will be protected from injury and sorrow if they drink water in which the flowers of the asoka have been soaked.
- Hindus also believe that the asoka is a symbol of love and dedicated to the Hindu God of Love, Kamadeva. They worship Kamadeva with garlands of red asoka blossoms.
- Its common name, *asoka*, is derived from Sanskrit and means 'without grief,' being thought to eliminate sorrow. Brahma, the God of Creation (another of the Hindu triad of gods), said: "He who eats eight buds of Asoka flowers on the eight day of the moon's increase in the month of Chaitra, marked by the asterism Punarvasu, suffers no bereavements in life".
- The asoka is also associated with human fertility and worshipped by women for offspring in ancient India.
- Being venerated, the asoka is commonly grown in the grounds of Buddhist and Hindu temples. This tree is planted in the southeast corner of the Hindu house or temple because it is sacred to Siva. Its leaves are used in religious ceremonies and its blossoms used as temple offerings.
- The initiation of the Jain Muni, Mahavira, attended by gods, was also under an asoka tree, so his attainment of omniscience is associated with this tree.
- Based on feng shui beliefs, the asoka, with its scarlet flowers, is symbolic of life, prosperity and the fire element, and so is auspicious for the front of the house, south, southwest or northeast zones of the garden; inauspicious for the west and northwest; and neutral for the rest.
- In northern Sri Lanka, the wood is used for building houses. Extract from the bark is used as an astringent for internal hemorrhoids and for treating uterine infections.

Solanum mammosum

apple of Sodom • nipple fruit

Scientific Species Name:
Solanum mammosum
(Latin *solanum*, probably referring to nightshade [*Solanum nigrum*]; Latin *mammosum*, furnished with nipples or breasts)

Common Species Name:
apple of Sodom, cow's udder, cow's udder plant, love apple, macaw bush, nipple fruit, pig's ears, titty fruit, zombi apple (English); *pomme zombi*, *tetons de jeune fille* (French); *Euter Nachtschatten*, *Zitzen Nachtschatten* (German); *rǔ qié* [乳茄] (Mandarin); *tetilla* (Spanish)

Scientific Family Name:
Solanaceae

Common Family Name:
nightshade family

Natural Distribution:
Mexico, Central America and Northern and Western South America and the Caribbean

USDA ≥10b (AUS ≥5)

How to Grow: Tropical shrub; most well-drained soils; propagated by seed or stem cuttings.

Drawbacks: Prickly stems and branches and spiny leaves; poisonous fruit; untidy-looking plant when older and unpruned.

Plant Description:

- The nipple plant is a small shrubby annual or perennial that can grow up to a height of 1.5 m (4.9 ft). It has flexible, hairy stems and branches, prickly throughout their length.
- The alternate, stalked leaves are large. The 6–15 cm (2.4–5.9 in) long leaf blades are broadly egg-shaped to heart-shaped, lobed with toothed margins. The entire leaf is hairy with long spines growing from the leaf veins.
- Flowers grow singly or in clusters of up to 6. They have purplish-blue petals that are joined into a short tube and 5 free tips.
- The yellow or orange 3–9 cm (1.2–3.5 in) long fruits are smooth, rounded, inverted and pear-shaped with 3 or 5 nipple-like protrusions near the stalk and contain many seeds.

 Photos: (opposite top) Melvin Bagaforo; (opposite bottom) Dave Quitoriano

Solanum mammosum leafy branches and fruits

Fruits with nipple-like protrusions

Folklore, Beliefs, Uses and Other Interesting Facts

- The fruit of the nipple plant is used by the Chinese as an auspicious table-top ornament during the Chinese New Year period. The fruits are often stacked into a tower. The Chinese believe that the fruit is symbolic of woman's sexuality and fertility, so by placing the fruit indoors, a young couple can be blessed with fertility and improve their chances of having a child in the coming year.
- The bright orange colour of the fruit is congruent with the lucky red hues of Chinese New Year and therefore thought to bring luck and fortune to its owner.
- According to feng shui beliefs, plants with thorns, prickles or spines are considered inauspicious, as the sharp structures represent 'poison arrows' which emanate hostile, inauspicious *qì* (energy or life force). However, because of the auspiciousness of the plant, it can be used as a sentry at the front gate. However, it should be placed as far away from the house as possible.
- In Central America, the seeds and leaves are used for treating colds and kidney ailments respectively.

Thuja occidentalis

white cedar • American *arbor-vitae*

Scientific Species Name:
Thuja occidentalis
(Greek *thuia*, a kind of juniper [*Juniperus* species]; Latin *occidentalis*, Western or of the Occident)

Common Species Names:
arbor-vitae, American *arbor-vitae*, Eastern *arbor-vitae*, Eastern white cedar, Northern white cedar, white cedar (English); *thuya d'Occident*, *thuya du Canada* (French); *Abendländische Lebensbaum*, *Abendländische Thuja* (German); *běi měi xiāng bǎi* [北美香柏] (Mandarin)

Scientific Family Name:
Cupressaceae

Common Family Name:
cypress family

Natural Distribution:
Eastern Canada and Northeastern USA

USDA 2–7 (AUS 1)

How to Grow: A slow-growing, low-maintenance, temperate, evergreen tree; most soil types that are medium moisture, well-drained, but prefers moist, neutral to alkaline, well-drained loams (especially of some limestone content); drought-intolerant; avoid windy sites; propagation by cuttings or seeds.

Drawbacks: Diseases include leaf blight, canker; pests include leaf miners, bagworms, mealybug, scales and spider mites; winter burn (turns yellow-brown) of foliage in exposed sites; may have stem/branch breakage in winter from ice and snow accumulations.

Plant Description:

- An evergreen, coniferous tree that grows to 15–38 m (49.2–124.7 ft) tall. The trunk has red-brown or grey-brown, fissured bark.
- The crown is conical. Branches are flat sprays that are perpendicular to the crown's surface.
- The opposite, densely arranged, 4-ranked, dull yellow-green scale leaves are 2 types; boat-like lateral ones which overlap the diamond-shaped facial ones that are 1.5–5 mm (0.06–0.20 in) long, and bear a distinct gland below.
- Each tree bears male and female cones. Male (pollen bearing) cones are reddish and 1–2 mm (0.04–0.08 in) long. Female (seed bearing) cones are ellipsoid, brown, 6–14 mm (0.24–0.55 in) long. Usually only about 2 pairs of cone scales bear seeds.
- The winged seed is red-brown, 4–7 mm (0.16–0.28 in) long including wings.

Thuja occidentalis seeds

Photos: (above) Steve Hurst @ USDA-NRCS PLANTS Database; (opposite top) USDA-NRCS PLANTS Database; (opposite bottom) K Jane Burpee

Thuja occidentalis tree

Cones, leaves and branches

Folklore, Beliefs, Uses and Other Interesting Facts

- The name *arbor-vitae* or *arborvitae* (Latin for tree of life) applies to *Platycladus* or *Thuja* species. There are five *Thuja* species: *Thuja koraiensis* (Korean *arbor-vitae*), *Thuja occidentalis* (American *arbor-vitae*), *Thuja plicata* (giant or western *arbor-vitae*), *Thuja standishii* (Japanese *arbor-vitae*) and *Thuja sutchuenensis* (Sichuan *arbor-vitae*). The two most commonly cultivated *arbor-vitae* species are *Platycladus orientalis* and *Thuja occidentalis*.
- In the winter of 1535–1536, the French explorer Jacques Cartier (1491–1557) and his men sailed up the St Lawrence River in North America for the northwest passage to China. Their fleet of three ships was frozen at a small fort (which later became Quebec City). Without fresh fruits and vegetables, they soon succumbed to scurvy, a disease caused by insufficient vitamin C. The native Huron Indians taught them to make a concoction from the bark and leaves of *arbor-vitae* and many of his men were saved. Cartier brought the plant back to France to share this great discovery. The then King of France, Francis I (1494–1547), declared it "*l'arbor de vie*", or "tree of life", hence its common name.
- North American Indians use *arbor-vitae* medicinally and ritually wherever it grows. Its branches are used to make teas to treat menstrual complaints, prostate problems and fever. Decoctions of the branches are used to induce abortion. The aromatic twigs are used for ritual incenses, e.g., in sweat baths. The twigs are also said to protect against magic.
- In Europe, the tips of branches and small leaves are boiled and used as a suforific, diuretic and expectorant in folk medicine.
- Magical superstition regarding junipers was also applied to the naturalised *arbor-vitae* in Europe.
- In the language of flowers, the *arbor-vitae* symbolises old age, or says "Live for me".
- Based on feng shui beliefs, the green and bushy form of the plant makes it symbolic of the wood element, so this plant is most auspicious for planting in the east, southeast or south zones of the garden, but inauspicious for the northeast and southwest, and neutral for the rest.
- This species is grown for its timber and medicinal oil, as a screen or windbreak, and as an ornamental (many cultivars, including variegated and dwarf ones), and also used in folk medicine.
- *Thuja occidentalis* is distinguished from *Platycladus orientalis* in having dry female cone scales when the cone is immature (versus fleshy female cone scales in *Platycladus orientalis*) and winged seeds (versus wingless seeds).

Thysanolaena latifolia

Asian broom grass • bamboo grass

Scientific Species Name:
Thysanolaena latifolia
(= *Thysanolaena agrostis*, *Thysanolaena maxima*) (Greek *thysanos*, fringe, and Greek *laina*, mantle or covering, possibly referring to the submarginally hairy leaf sheaths; Latin *latifolia*, broad-leafed, referring to the wide leaf blades relative to those of most other grasses)

Common Species Names:
Asian broom grass, bamboo grass, bouquet grass; kuntz, tiger grass (English); *amliso*, *herbe à balai* (French); *buluh rumput*, *buluh tebrau* (Malay); *dà zòng yè lú* [大粽叶芦 / 大粽葉蘆], *jiàn zhú máo* [箭竹茅] (Mandarin)

Scientific Family Name:
Gramineae or **Poaceae**

Common Family Name:
grass family

Natural Distribution:
India to China and southwards to Peninsular Malaysia (usually 300 m [984.3 feet] above sea level)

USDA ≥6 (AUS ≥1)

How to Grow: Temperate and highland tropical grass of open areas of the forest, riverbanks, grassland and hills; intolerant of overlong periods of frost; most soils, pH 5.3–9.3; fertilise regularly; propagate by stem or rhizome cuttings, dividing the clump of stems or seed.

Drawbacks: Not for small gardens; can get untidy with growth.

Plant Description:

- A leafy, bamboo-like herb of up to 4 m (13.1 ft) tall, growing from rhizomes.
- Its stems, unbranched, solid and up to 1.5 cm (0.6 in) wide, are joined at their bases, developing into a dense clump.
- Leaf-sheaths, about 12 cm (4.7 in) long, are tight, smooth, smooth and hairless, except at the sub-margins. The leaf blades, up to 60 cm (23.6 in) by 7 cm (2.8 in), are lance-shaped, smooth and hairless, slightly leathery and stalked. The lower leaf blade surface is covered with a waxy bloom.
- The tiny flowers are organized in a branching inflorescence at the stem tip, 20–60 cm (7.9–23.6 in) long.
- Fruits are tiny grains.

 Photos: (opposite) Hugh Tan Tiang Wah

Thysanolaena latifolia plant

Leafy branch tips

Inflorescence at shoot tip

Folklore, Beliefs, Uses and Other Interesting Facts

- According to Vajrayana Buddhism, the leaves of broom grass are sacred because it is believed that Gautama Buddha attained enlightenment while sitting on the leaves of broom grass under the bodhi tree (*Ficus religiosa*; refer to its own entry). Vajrayana Buddhists continue to use broom grass leaves as sitting mats during meditation and prayer sessions, as the leaf is seen as a spiritual tool that helps to clarify their minds during these moments of intense concentration.
- According to feng shui beliefs, the leafy, green appearance of broom grass will make it auspicious for the wood element zones of the garden such as the east, southeast and south; inauspicious for the southwest or northeast; and neutral for the rest.
- Sometimes cultivated as a large, ornamental grass, the plant's broom inflorescence was important for broom-making in Southeast Asia. The leaf blades are used to wrap Chinese glutinous rice dumplings, the young leaves and stem tips for fodder and, in the highlands of Java, broom grass is occasionally used as a fence or screen.
- This species is the only one in its genus.

Scientific Species Name:
Vicia faba
(Latin *vicia*, vetch [*Vicia* species]; Latin *faba*, broad bean)

Common Species Name:
broad bean, English bean, European bean, faba bean, fava bean, field bean, horse bean, Windsor bean (English); *fève*, *fève à cheval*, *fève de cheval*, *fève des marais*, *féverole*, *grosse fève*, *grosse fève commune*, *gorgane*, *gourgane* (French); *Ackerbohne*, *Dicke Bohne*, *Pferdebohne*, *Puffbohne*, *Saubohne*, *Saubohnen* (German); *anhuri*, *bakla*, *kala matar* (Hindi); *kacang babi* (Malay); *cán dòu* [蚕豆/蠶豆] (Mandarin); *faba*, *haba*, *haba caballar*, *haba común*, *haba de huerta*, *haba mayor*, *habichuela*, *haboncillo* (Spanish)

Scientific Family Name:
Fabaceae or **Leguminosae**
(formerly in the Papilionaceae)

Common Family Name:
bean family

Natural Distribution:
Thought to be a cultigen derived from *Vicia narbonensis* which naturally occurs in the Mediterranean to Central Asia

How to Grow: Temperate crop plant; prefers rich loams, with optimum pH of 6.5 but grows in pH 4.5–8.3; most cultivars are frost-intolerant; propagate by seed.

Drawbacks: Usually a long-day plant, meaning that it flowers as the nights get shorter (in summer), but some cultivars are day-neutral (will flower irregardless of day length); usually fruitless in the tropics; with numerous fungal, viral, nematode and bacterial diseases; pests include the faba bean weevil, bean aphid; dangerous to eat raw for some people (refer to Folklore, Beliefs, Uses and Other Interesting Facts).

Plant Description:

- A stiffly erect, annual herb, 30–80 cm (11.8–31.5 in) tall.
- The thick, hollow stem has a square cross section and branches only at the base.
- The alternate, pinnate leaves have 2–6 leaflets, each up to 10 cm (3.9 in) long and 4 cm (1.6 in) wide.
- The axillary inflorescence is an elongated shoot of 1–6 fragrant flowers each. The petals are white streaked or blotched black or purple, or red.
- The fruit pod is somewhat flattened, up to 30 cm (11.8 in) long, containing variably shaped and sized seeds which may be white, green, yellowish brown, purple or black.

 Photos: (opposite top) Pat Knight; (opposite bottom) Rowena Wood

Vicia faba flowers on a leafy shoot

Vicia faba 'Masterpiece Green Longpod' as a garden plant in Bath, UK

Folklore, Beliefs, Uses and Other Interesting Facts

- In the British Isles, up to the end of the 19th century, people believed that the souls of the deceased resided in the fava bean flower. This belief has persisted in that the flowers are still seen as inauspicious. When the flowers bloom, cases of insanity are thought to become more frequent, as the fragrance of the flower is believed to cause nightmares, frightening images and psychological disorders. A traditional superstition in Leicestershire has it that one will suffer from horrifying nightmares if one sleeps in a field of fava bean plants for a night. Another superstition is that if one plant in a row becomes white (displays albinism, lacking any chlorophyll), it is a sign that someone in the farmer's family will die. On the other hand, warts are thought to be cured by rubbing them with the inner surface of a fava bean pod, the warts disappearing as the discarded pod rots.
- Some people develop favism (break up of the red blood cells) when they consume raw fava beans. Such people lack the enzyme, glucose-6-phosphate dehydrogenase (G6PD), which is a rare genetic deficiency usually afflicting a minority of African Americans, Greeks, Italians, Chinese and others. As this is an X-chromosome linked gene, it is more frequent in males than females.
- Favism has been thought to be the reason why Pythagoras (around 580–500 BC), the ancient Greek philosopher, did not allow his followers to eat the bean and placed a ban on its consumption. However, this hypothesis has now been dismissed by recent research. More plausible explanations are the association of its black seeds with dying, hell and its forces, impurity and sadness. (The black seed coat was more typical of the plant in ancient Greece; the other colours were bred in more recent times.)
- Based on feng shui beliefs, the fava bean is not auspicious for the home garden, considering its white flowers with black streaks and blotches (white and black being the colour of mourning or death). If one were a diehard fan of the plant, then it is most auspicious to plant it in the west, northwest or north zones of the garden. It is inauspicious to plant in the front of the house, east or southeast, and neutral for the other zones.
- Presently, the fava bean is a major food crop for humans and animal feed.

 common mistletoe

Scientific Species Name:
Viscum album
(Latin *viscum*, mistletoe; Latin *album*, white, referring to its white berries)

Common Species Name:
common mistletoe, European mistletoe (English); *gui* (French); *Mistel* (German); *akasha beli, amara bela* (Hindi); *dedalu* (Malay); *bái guǒ hú jì shēng* [白果槲寄生], *luǎn yè hú jì shēng* [卵叶槲寄生/卵葉槲寄生] (Mandarin); *muérdago* (Spanish)

Scientific Family Name:
Santalaceae
(previously in the Viscaceae)

Common Family Name:
sandalwood family
(previously in the mistletoe family)

Natural Distribution:
Europe, North Africa, North and West Asia, Afghanistan, Pakistan, India and Japan

USDA 6–8 (AUS 1–2)

How to Grow: This is a natural parasite growing on other plants.There are three subspecies; two subspecies infect only conifers in Central and Southern Europe and Northern Turkey, the last subspecies infects trees such as apple (mostly hybrids of *Malus* species), hawthorn (*Crataegus* species), lime (*Tilia* species), poplar (*Populus* species), etc., by growing into their branches to tap their water supply directly from the vascular tissues. This parasitic plant can even parasitise another plant parasite, *Loranthus europaeus*. It propagates by the seed; birds eat the fruits but the viscid tissue surrounding the seed prevents it being swallowed and the bird scrapes it off onto the bark of the host tree, where it germinates, and grows into the host. To artificially infect trees, seeds can be scraped onto bark of twigs.

Drawbacks: Mistletoe infected trees can weaken and die; all parts of the plant are poisonous.

Plant Description:

- Evergreen, perennial, unisexual shrub to 1 m (3.3 ft) tall, parasitic on trees or another mistletoe (*Loranthus europaeus*).
- Branches are forked and hang downwards.
- The opposite, stalkless leaves are 2.5–7 cm (1–2.8 in) long, with leathery, elliptic, oval or egg-shaped leaf blades with a smooth margin.
- The tiny, unisexual, stalkless, yellow flowers develop in the axils of the leaves in clusters of 3–5. Each flower has 3–4 tepals. Male flowers possess 4 stamens. Female flowers contain one egg-shaped ovary topped by a short conical stigma.
- The fruit is a white, translucent, round berry containing one seed embedded in a sticky pulp.

 Photos: (opposite top) Wlodzimierz Lukasik; (opposite middle and bottom) Peter M Forster

Viscum album male flowers

Fruiting and leafy branches

Viscum album plants on beech tree host

Folklore, Beliefs, Uses and Other Interesting Facts

- Magical powers and myths have been attributed to this unusual plant that grows on other trees, stays green and leafy even in winter and possesses forked branches. In Greek mythology, Persephone used a magical wand of mistletoe to open the gates of the underworld. In Scandinavian mythology, the god Baldur was killed by a mistletoe branch, the only thing able to hurt him. Pliny the Elder (AD 23–79) recorded beliefs about mistletoes, including fertility advancement, defence against poisons and that the mistletoe is considered the most sacred of all plants to the Gallic druids. However, such pagan associations have led this plant to be traditionally banned in churches in the British Isles.
- The mistletoe is associated with Christmas, especially in the northern hemisphere. It symbolises undying life during winter, when most plant life is dormant, and is used in magical rites to ensure the return of plant growth. Sacred buildings in Europe and Western Asia were decorated with Christmas evergreens during rituals of the winter solstice.
- The tradition of associating mistletoe with Christmas since the mid-19th century may also have stemmed from practical reasons; besides this species, evergreens such as the holly (*Ilex aquilifolium*) and ivy (*Hedera helix*) were the only plants available in winter in the northern hemisphere. They were used to add colour to homes. Sprigs of mistletoe are sold commercially for this use.
- Like the holly, the presence of mistletoe in the house (at any time other than Christmas) is deemed to be unlucky. Therefore, mistletoe should also be brought into the home only on Christmas Eve and removed before the 12th day of Christmas.
- The mistletoe is regarded as a plant of love and peace under which couples kiss. Hence the popular phrase: "Kissing under a mistletoe tree". This is, apparently, an English custom, perhaps associated with the belief that paired fruits are a fertility symbol.
- In the language of flowers, the mistletoe indicates obstacles to overcome or says "I surmount all. "
- The plant's polypeptides (lectins) are immune-system stimulating and poisonous to mammals but not to birds, the dispersers of its seeds.

Vitis vinifera grape • common grapevine

Scientific Species Name:
Vitis vinifera
(Latin *vitis*, the grape; Latin *vinifera*, wine-producing)

Common Species Name:
common grapevine, European grape, grape, grapevine, vine (English); *raisin, grappe, vigne, vigne cultivée* (French); *echter Weinstock, Rebe, Rebstock,* Wein, Weinrebe, Wein-Rebe, *Weinstock* (German); *anggur, buah anggur* (Malay); *ōu zhōu pú táo* [欧洲葡萄/歐洲葡萄], *pú táo* [葡萄] (Mandarin); *vino* (Spanish)

Scientific Family Name:
Vitaceae

Common Family Name:
grape family

Natural Distribution:
Northeast Afghanistan to the Southern borders of the Black Sea; now widely cultivated to latitude 52°N in Poland; cultivars of this species and hybrids with other *Vitis* species now grown

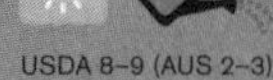

USDA 8–9 (AUS 2–3)

How to Grow: Fast-growing Mediterranean climber but can be grown even in the tropics with care; can tolerate semi-shaded conditions; acid to slightly acidic [pH 5–8, just below 7 is best], medium-draining soils of any texture; high drought tolerance; flooding intolerant; variable salt tolerance depending on the rootstock used; propagated by stem cuttings and grafts on suitable rootstocks; seeds do not germinate unless they are cold treated but seedlings are not true to type.

Drawbacks: Downy mildew, anthracnose diseases (fungal); virus diseases; cutworms, flea beetles, leafhoppers, thrips, red spider mites, nematodes, grape root louse, etc., are its pests; fruits attract bats and birds; pluck fruits only when they ripen on the vine as they do not ripen after harvesting.

Plant Description:

- Deciduous, perennial, woody climber (vine) growing to ≥20 m (65.6 ft) long or more. In cultivation, often reduced in size by pruning.
- The stalked leaves are alternate with usually forked tendrils opposite them. Leaf blades are circular to circular-oval in outline with a few broad finger-like lobes and broadly toothed margins.
- The fragrant, tiny, functionally or actually unisexual flowers occur in dense, many-flowered clusters opposite the leaves. There are 5 petals, growing up to 5 mm (0.2 in) long, whose tips are joined into a cap which drops off.
- The fruit, in bunches up to several, is a juicy berry, round to oblong, to 25 mm (1.0 in) long, variably coloured depending on the cultivated variety—dark bluish purple, red, green or yellow. The flesh is sweet or sour.
- Each fruit contains 0 (seedless cultivars) to 3–4, pear-shaped seeds.

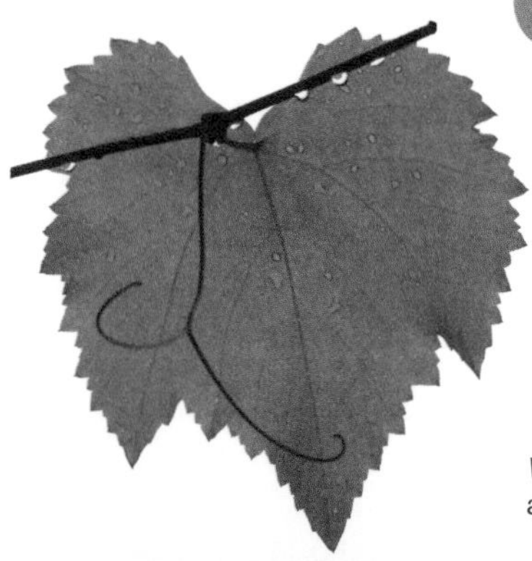

Vitis vinifera leaf on a stem with a forked tendril opposite it

Photos: (above and opposite top) Hugh Tan Tiang Wah; (opposite bottom) Bob Nichols, ARS-USDA

Vitis vinifera cultivar fruit bunches on leafy vines

Crimson seedless table grapes

Folklore, Beliefs, Uses and Other Interesting Facts

- In Deuteronomy 8:7–8 (King James Version), the people of Israel were promised—"....a good land....A land of wheat, and barley, and vines, and fig trees, and pomegranates; a land of oil olive, and honey," so these have a special place in Judaism as the seven plants of the Bible. 'Grape' and its related terminology (e.g., vine, vineyard) are mentioned more than 200 times in the Bible (King James Version)!
- In terms of cultivated area, production and value, the grape is the world's largest fruit crop.
- Most grape production is for its juice to be made into wine, second for fresh grape fruits (table grapes) and, thirdly, for dried grapes (raisins, sultanas and currants). Grape juice is also important. Brandy or alcohol are added to wines to produce port, madeira, sherry, etc., and, with flavourings added, become vermouths and martinis, and with quinine, iron and others, made medicinal.
- Grape seed oil is used for cooking and in cosmetics.
- Oligomeric procyanidins are extracted from grape seed which is a major source. This substance is also found in greater quantities in red wine compared to white, and implicated as being the cause for reduced risk of coronary heart disease and lowering of the overall mortality amongst regular and moderate consumers of red wine.

Zamioculcas zamiifolia

zz plant • money tree

Scientific Species Name:
Zamioculcas zamiifolia
(*Zamioculcas* derived from the merger of names, *Zamia* and *Culcas*, the former derived from Latin *zamiae*, a corruption of *azaniae*, pine cones, and the latter, from the vernacular name; *zamiifolia*, after the genus *Zamia,* a generic name for a cycad, and *-folius* in Latin compound words means -leafed, hence *Zamia*-leafed)

Common Species Name:
aroid palm, eternity plant, fat boy, money tree, zz plant (English); *plante zz* (French); *jīn qián shù* [金钱树/金錢樹] (Mandarin)

Scientific Family Name:
Araceae

Common Family Name:
aroid family

Natural Distribution:
East tropical Africa (from Kenya to northeast South Africa)

USDA ≥11 (AUS ≥4)

How to Grow: Tropical indoor plant; very easy to grow because it is drought-resistant, tolerant of low humidity, low light levels and generally pest-free; well-drained soil (natural habitat is rocky ground of dry lowland forest); fertilise sparingly with half or quarter recommended dose as the plant is slow-growing; propagation by seed or leaf stalk or leaf axis pieces, or placing whole or parts of leaflets in rooting medium.

Drawbacks: Prone to fungal rot of the leaf stalks and death from excessive watering or permanent wetness.

Plant Description:

- A slow-growing, perennial herb bearing alternate, fleshy stalked, pinnate leaves growing to about 80 cm (31.5 in) long, extending from a stout underground rhizome.
- Leaflets are very shiny, dark-green above, leathery and oval-shaped.
- The inflorescence is a yellow-green to bronze spadix 5–7 cm (2.0–2.8 in) long.

Zamioculcas zamiifolia inflorescence

 Photos: (above and opposite) Hugh Tan Tiang Wah

Zamioculcas zamiifolia plants for sale in a nursery during Chinese New Year

Plants adorned with Chinese New Year ornaments for prosperity and good luck

Folklore, Beliefs, Uses and Other Interesting Facts

- The zz plant is also known as the eternity plant because of its hardiness under the most extreme indoor conditions, except over-watering. It is usually an evergreen but becomes deciduous in its natural habitat to survive drought periods. The leaves are shed, and the plant survives on the stored food in the rhizomes until the drought is over. It is grown as an indoor leaf ornamental for its beauty.
- Among the Chinese in Asia, the zz plant is also known as *jīn qián shù* (Mandarin for 'gold coin plant') because the shiny deep-green leaflets arranged on the rachis resemble the hollow-centred coins which were strung together in ancient China. This auspicious plant symbolises the possession of abundant money and is believed to bring fortune to its owner. It is commonly sold during the Chinese New Year and decorated with red ribbons or gold-coloured ornaments for display in the home.
- This species is the only one in the genus *Zamioculcas*.

Zantedeschia aethiopica arum lily

Scientific Species Name:
Zantedeschia aethiopica
(= *Calla aeithiopica*)
(Probably after Dr. Giovanni Zantedeschi (1773–1846), an Italian physician and botanist from Verona; Latin *aeithiopica*, of or from Ethiopia; Greek *kallos*, beauty)

Common Species Name:
arum lily, calla lily, common arum lily, common calla, common calla lily, Egyptian lily, florist's calla, garden calla, giant white arum lily, Jack in the pulpit, lily of the Nile, pig lily, trumpet lily, white arum lily (English); *arum* (French); *mǎ tí lián* [马蹄莲/馬蹄蓮] (Mandarin)

Scientific Family Name:
Araceae

Common Family Name:
aroid family

Natural Distribution:
Southern Africa; naturalised in Australia, Macronesia, Mascarenes, New Zealand, UK and USA (California, Hawaii and Puerto Rico)

USDA 8–10 (AUS 2–4)

How to Grow: Rich, organic soil, moist soil, or even submerged under shallow water; afternoon shade required for hotter climates; propagated by dividing the rhizome or seed.

Drawbacks: All plant parts are poisonous and sap may irritate skin.

Plant Description:

- An evergreen, rhizome-bearing, herbaceous plant that grows to about 1.5 m (4.9 ft) tall.
- The stalked leaves arise from the underground rhizome, and their shiny, dark green leaf blades are 15–20 cm (5.9–7.9 in) by 10–15 cm (3.9–5.9 in) long and oval to arrowhead-shaped.
- The bright yellow spike-like inflorescence (spadix) rises on a long vertical stalk and is surrounded by a large ivory-white modified leaf called a spathe. The spadix consist of tiny stalkless male and female flowers. Female flowers consist of green ovaries surrounded by white staminodia (sterile stamens).

Zantedeschia aethiopica inflorescences and plants

Zantedeschia aethiopica inflorescence

Folklore, Beliefs, Uses and Other Interesting Facts

- In many countries, the arum lily is usually considered unsuitable for home floral arrangements or hospitals because is it is used for mourning and decorating graves.
- Based on feng shui beliefs, this species, with a white spathe surrounding the inflorescence (or any white-flowered species), would be associated with mourning and death, so it would not be planted in home gardens. For fans of its beauty, it is best planted in the metal element zones of the garden—good for the west, northwest and north, bad for the house front, east and southeast, and neutral for the rest.
- The Easter Lily is another name for the arum lily. It is associated with the 1916 Easter Rising by the Irish at Dublin, in an attempt to oust British Rule.
- In the language of flowers, the arum lily signifies return of happiness or delicacy.
- It is the birth month flower of May, signifying the return of happiness since it flowers in spring.
- In ancient Egypt, during the New Kingdom dynasty (1570–1085 BC), together with other flowers, the arum lily was used for offerings at the religious altar.
- This attractive species has become an invasive weed in countries, as the seeds are spread by birds and the dumping of cuttings from cultivated plants. Hawaii, La Réunion Island and New Zealand are some of the afflicted territories.
- Arum lily inflorescences and leaves make lasting floral arrangements. The attractive plants are grown outdoors in ornamental ponds or as house or greenhouse plants.
- The arum lily is neither an arum (*Arum* species) nor a lily (any member of the lily family or Liliaceae).

Glossary, References and Acknowledgements

alternate: Of leaves; with only one leaf inserted at each node (joint) of the stem, e.g., peach.

air layering: A method of propagation whereby a portion of a woody twig's bark is ringed, or an inclined cut made halfway into it; this region wrapped in a moist medium like sphagnum moss and all of these sealed inside a polythene bag or other suitable container. Over time, the wounded areas will produce roots and the twig ultimately cut below the rooting zone to be grown as another plant.

aril: Outgrowth of the seed stalk which grows to cover the seed; usually brightly coloured and/or fleshy to attract animal dispersers.

axil: This is the angle made by the upper side of the base of the leaf stalk (for a stalked leaf) or the leaf blade (for a stalkless leaf) with the twig or branch to which it is attached. One or more buds are usually found here and the buds may develop into another branch, flower or inflorescence.

axillary: Pertaining to the axil. Structures developing at the axil are qualified with this adjective, e.g., axillary bud, axillary inflorescence.

berry: A fleshy fruit that contains one or more seeds and does not split open when ripe.

bipinnate: Of leaves; twice-pinnate.

bud grafting: A method of propagation where a bud and a portion of the stem to which it was attached from one plant is inserted into the bark of the stem of another so that the bud develops after fusion between the two.

bulb: A shortened stem enclosed by concentric fleshy leaf bases, e.g., onion.

calyx: The collective name for all the sepals in a flower when they are separate, and also the compound structure composed of the sepals when they are joined together.

capitulum: A kind of inflorescence where the flowers are clustered into a round structure (inserted at the tip of a stalk), or a flattened structure (inserted on a flat disk), e.g., daisy, sunflower.

capsule: A dry, two- or more-loculate fruit which splits when ripe to release it seeds.

carpel: The organ which bears the ovules (seed precursors); a simple pistil (one-loculate) or one component of a compound pistil.

caryopsis: A dry fruit where the fruit wall is fused to the seed coat of the single seed within, e.g., grasses.

catkin: A spike that hangs downwards rather than standing erect, e.g., birch.

climber: A woody or non-woody plant whose stem is weak and unable to stand erect on its own and climbs on other plants or structures to gain height to reach sunlight. When growing in open, level areas, climbers grow flat or trail on the ground (trailer).

cone: The reproductive structure of a gymnosperm (e.g., ginkgo, conifers) that usually consists of an axis to which are attached, spirally arranged structures that bear pollen or seeds.

conifer: Woody plant that possesses water- and food-conducting tissues and produces leaves, cones and seeds, e.g., pine.

corolla: The collective name for all the petals in a flower when they are separate, and also the compound structure composed of all the petals when they are joined together.

crown: The branches and foliage of a tree at the upper portion of the trunk.

culm: The stem of a grass (any member of the grass family, the *Poaceae* or *Gramineae*), including the bamboos.

cultigen: Refers to a species whose origin is undetermined because there is no known wild type; also refers to a species that has been cultivated for such a long time and has transformed so much that its ancestry is unknown.

cultivar: The short form for cultivated variety.

cultivated variety All the plant individuals propagated from a single individual by vegetative propagation such as stem cuttings, grafting, air layering, or by tissue culture techniques. Theoretically, all the propagated plants have the same genetic makeup as the original plant from which they were propagated.

cuttings: Stem cuttings.

deciduous: A plant which sheds its leaves for part of the year (winter, dry season), e.g., temperate, dicotyledonous trees.

dehisce: Split.

dehiscent: Splitting.

dicotyledon: One of the main kinds of flowering plant, characterised by having seedlings with usually two seed leaves (cotyledons), leaves with net-like veins and other features.

dicotyledonous: Pertaining to dicotyledons.

disc floret: Inflorescences like those of the sunflower or daisy are composed of two kinds of florets. Those which resemble large petals at the periphery are the ray florets; those in the centre, are the disc florets.

double, double flower: A flower with many more petals than the wild type condition, the extra 'petals' usually derived from stamens which have become petaloid (petal-like) through selective breeding.

endocarp: The innermost layer of the fruit wall; may be fleshy (e.g., berry) or hard and tough (e.g., drupe).

epicarp: The outermost layer of the fruit wall; the 'skin'; exocarp.

evergreen: A plant which bears it leaves throughout the year, e.g., conifers, most tropical plants.

drupe: A fruit where the innermost layer of the fruit wall, the endocarp, is hard and tough, surrounding one or more seeds, whereas the outer two layers, the epicarp and mesocarp are fleshy, e.g., mango, plum.

feng shui: The Chinese practice of geomancy with ancient roots to achieve harmony with the environment; used for siting of houses, buildings, burial sites, designing homes, etc. (*fēng shuǐ*; 風水; literally "wind water").

fern: Usually herbaceous plant that possesses water- and food-conducting tissues and produces vegetative and spore-producing leaves but no seeds.

floret: A small flower.

flower: Reproductive structure consisting typically, from the base upwards, of a flower stalk, receptacle to which the sepals, petals, stamens and pistil(s) are attached.

flowering plant Herbaceous to woody plant that possesses water- and food-conducting tissues and produces leaves, flowers, fruits and seeds; also called an angiosperm.

follicle: A dry, one-loculate fruit which splits along one side when ripe to release its seed(s).

glabrous: A surface which is smooth and hairless.

grain: A caryopsis.

head: A capitulum.

herb: A non-woody, usually small, plant.

herbaceous: Non-woody.

hybrid: The offspring of the breeding between two different species; annotated by a multiplication sign before the generic name for hybrids of two species belonging to different genera, e.g., x*Citrofortunella microcarpa* (limequat), or the specific epithet for hybrids of two different species from the same genus, e.g., *Prunus* x*yedoensis* (cherry blossom).

imparipinnate: A pinnate leaf where there is a terminal leaflet, and the lower leaflets are opposite each other along the long axis of the leaf.

inflorescence: A flowering shoot that consists of two or more flowers. A simple inflorescence is an unbranched structure, and a compound inflorescence, one with more branches.

juvenile: Pertaining to structures associated with the young plant.

lanceolate: Lance-shaped.

layering: A method of propagation where a branch is bent down into the soil by pegging and the branch rooting where it contacts the soil while still attached to the plant. When the rooting is sufficient, the branch is cut from the parent plant and grown as a separate plant.

legume: A dry, one-loculate fruit which splits along two sides when ripe, to release its seed(s).

linear: Long and slender like a line.

locule: A space or cavity within a structure.

long shoot: In some plants, the branches consist of short (spur) and long shoots, e.g., pine, ginkgo. Long shoots grow quickly and indefinitely so their leaves tend to grow widely apart because the nodes (joints) to which they are inserted are widely separated. Short shoots develop from the axillary buds of the long shoots.

mesocarp: The middle of the three layers of the fruit wall, usually fleshy.

monocotyledon: One of the main kinds of flowering plant, characterised by having seedlings with usually one seed leaf (cotyledon), usually leaves with parallel veins, and other features.

mulch: Organic or inorganic matter used to cover the soil to prevent excessive evaporation and soil erosion; to cover with mulch.

needle leaf: A leaf which is long and narrow, like a needle, e.g., pine.

nut: A dry, hard-coated fruit which does not split when ripe.

nutlet: A small nut.

oblanceolate: Lance-shaped but attached by the narrower end.

obovate: Egg-shaped but attached by the narrower end.

opposite: Of leaves—with two leaves inserted at each node (joint) of the stem or branch, such that they are opposite each other.

ovate: Egg-shaped and attached by the broader end.

ovatelanceolate: Intermediate between egg- and lance-shaped.

ovoid: Egg-shaped solid.

palmate, palmately compound: Compound leaf where the leaflets are inserted at the tip of the leaf stalk, like the fingers of the palm, e.g., kapok tree.

perennial: A plant which lives for more than two years, usually much longer, e.g., trees.

petaloid: Resembling a petal.

pH: A unit of measure of the acidity or alkalinity of a solution. Neutral is pH 7; below pH 7 is acidic and the smaller the number the more acidic; above pH 7 is alkaline, and the larger the number the more alkaline.

pinnate: A kind of compound leaf where the leaf blade is divided into leaflets that are attached to the long axis of the leaf, like a feather.

pistil: The female part of the flower, typically consisting, from the base upwards, of the ovary, style and stigma, the site for receiving pollen grains. The pistil may be simple, consisting only of one carpel, or compound, being made up of two or more carpels.

pome: A fleshy fruit like that of the apple.

pontianak: A female vampire of Malay folklore and who terrorises the living; originally a woman who died at childbirth and became undead.

prickle: A hard, sharp outgrowths of the plant's stem surface but not developing from the leaf axil, e.g., rose stems.

pyrene: The stone or the hard innermost fruit wall surrounding the seed(s) in a drupe. A drupe may have one (e.g., plum) or more (e.g., coffee) pyrenes within.

qì: Based on feng shui beliefs, this is energy, life force or vital breath; can also mean air, gas, smell, weather (气/氣).

ray floret: Inflorescences like those of the sunflower or daisy are composed of two kinds of florets. Those which resemble petals at the periphery are the ray florets; those in the centre, are the disc florets.

rhizome: A stem which grows horizontally at or below the soil surface, e.g., ginger.

rhombic: Diamond-shaped.

samara: A dry, winged fruit that does not split open when ripe, e.g., birch.

scale leaf: A small leaf which grows flat on or close to the stem and/or overlapping with others.

semi-double, semi-double flower: Having slightly more petals than the wild type condition, some of the extra 'petals' having derived usually from stamens that have become petaloid.

short shoot: In some plants, the branches consist of short (spur) and long shoots, e.g., pine, ginkgo. Short shoots grow slowly and then stop their growth so their leaves tend to grow closely together because the nodes (joints) to which they are inserted are close together. Short shoots develop from the axillary buds of the long shoots.

shrub: A woody, multiple-trunked, usually intermediate to short plant.

spike: An erect, unbranched flowering shoot which bears stalkless flowers that develop from the bottom upwards consecutively.

spikelet: A small spike, e.g., grasses.

spadix: A spike which is enveloped or subtended by a large modified leaf called a spathe, e.g., aroids.

spine: A hard, sharp outgrowths of the plant's leaves, including the stipules, leaf stalk, leaf sheath, leaf blade, leaflets and so on, e.g., holly leaves.

spirally arranged: Of leaves. With the leaves inserted only one per node (joint) of the stem and with the leaves appearing to form a spiral when viewed from the tip of the stem, e.g., pineapple leaves.

stamen: The male organ of the flower, typically consisting of the filament and anther that produces the pollen.

stem cutting: A method of propagation where a part of a stem or branch cut from the main plant which used for rooting by placing the lower cut end into a suitable medium, in order to produce another plant.

stipule: Appendage of a leaf, typically at the base of a leaf stalk, resembling a tiny leaf blade.

stipulate: Possessing one or two stipules.

subtropical: Pertaining to the subtropics.

sucker: A shoot which arises from the underground part of a stem or root; a natural means of vegetative propagation for a plant, e.g., pineapple.

subtropics: Regions which are immediately north or south of the tropics.

temperate: Pertaining to the regions between the tropics and polar circles (approximately latitudes 23° 27' to 66° 33' N or S) and characterised by non-extreme, seasonal climates.

tendril: A coiled structure used by plants for attaching to supports to enable them to climb.

tepal: In some flowers, the flower parts are not distinguished into the sepals (usually green and smaller) and petals (usually non-green, brightly coloured and larger), but consist only of one kind of part (green or colourful). (It is not a typographical error for petal!)

terminal: At the tip of a structure such as a stem.

The Four Gentlemen of Flowers: Based on traditions of ancient Chinese painting, the four seasons are represented by these plants: (1) spring/orchid; (2) summer/bamboo; (3) autumn/chrysanthemum; (4) winter/plum.

thorn: A hard, sharp outgrowth that develops from a bud at the axil of the leaf, e.g., citruses.

tree: A woody, single-trunked, usually large, plant.

tripinnate: Thrice pinnate.

tropical: Pertaining to the tropics

tropics: The area on the earth's surface approximately between latitudes 23° 27'N and S (Tropic of Cancer and Tropic of Capricorn, respectively), and characterised by regions with climates that are frost-free, wet, humid and warm throughout the year.

variegated: With marks or patches of a different colour on leaves or other structures.

variety: A botanical variety which is a subset of a species; similar to a subspecies.

varietal: Pertaining to a botanical variety.

yin: Based on feng shui belief, the dark element, feminine, overcast or cloudy, shady, negative, moon (*yīn*; 阴/陰).

yang: Based on feng shui belief, the bright element, masculine, sunny, bright, positive, sun (*yáng*; 阳/陽).

The references for the individual plant descriptions follow, with the details of each given in detail in the alphabetical list at the end of this section.

The following references were relied upon for many of the descriptions, so are not indicated for the individual plant species descriptions, in particular, for the etymology of the names: Backer (1936), Jaeger (1950), Chiceley Plowden (1968), Stearn (1996); for general botanical information: Griffiths (1992), Mabberley (1997), USDA, ARS, National Genetic Resources Program (2007); for horticultural information of tropical plants: Boo, Kartini Omar-Hor and Ou-Yang (2006); for feng shui information: Too (2002).

Plant Description References

Plant descriptions and their reference sources are here:

	Plant Entry	References
1	*Adenium obesum*	Tan (1991); Faucon (1998–2005)
2	*Aegle marmelos*	Gupta (1971); Morton (1987); Parma and Kaushal (1982)
3	*Alstonia scholaris*	Burkill (1966); Corner (1983); Rao and Wee (1989)
4	*Ananas comosus*	Duke (1983); Morton (1987)
5	*Azadirachta indica*	Gupta (1971); National Research Council (1992)
6	Bambuseae	Sze (1963); Goody (1993); Chua, Soong and Tan (1996); Too (1999); Wong (2004)
7	*Betula* species	Radford and Radford (1961); Upadhyaya (1964); Gilman and Watson (1993a–d); Vuokko (2001); Kendall (2006); Wichman and co-authors (2006); Furlow (Undated)
8	*Camellia japonica*	Seaton (1995); Christman (1997a); Gilman (1999a); Enbutsu (2002); Wen (2004 onwards); The United States National Arboretum (2006); Chongqing Municipal Government (2007); The International Camellia Society (2007); Matsuyama City Office (Undated)
9	*Cananga odorata*	Burkill (1966); Corner (1983)
10	*Canavalia gladiata*	Kooi (1994); Bosch (2004), BBC News (2005), Wong (2007a–b)
11	*Cardamine pratensis*	Radford and Radford (1961); Zhou and co-authors (2001); Darlington & Stockton Times (2006); Healey Dell Nature Reserve (2006)
12	*Castanospermum australe*	Roja and Heble (1995); Debevec (2002)
13	*Ceiba pentandra*	Burkill (1966)
14	*Cercis siliquastrum*	Robertson and Lee (1976); Hora (1981); Ciesla (2002), Fraser (2002); Ali (Undated); Royal Horticultural Society (Undated)

	Plant Entry	References
15	Christmas Tree	Tille (1892); Chastagner and Benson (2000); Fagg (2002); Bates (2003); Earle (2004a–d, 2005, 2006a–c, 2007); Keen (2007); National Fire Protection Agency (2007); Bone (Undated); Chastagner and Hinesley (Undated); Nix (Undated)
16	*Cibotium barometz*	Burkill (1966); Holttum (1968)
17	x*Citrofortunella microcarpa*	Morton (1987); Sotto (1992)
18	*Citrus medica* var. *sarcodactylis*	Burkill (1966); Jones (1992); Gilman and Watson (1993e)
19	*Citrus reticulata*	Ashari (1992); Seaton (1995)
20	*Cocos nucifera*	Burkill (1966); Gupta (1971); Rao and Wee (1989); Gillman and Watson (1993f); Lai (2003)
21	*Conium maculatum*	Sievers (1930); Radford and Radford (1961); Rätsch (1992); Seaton (1995); Dave's Garden (2000–2007c); Pan and Watson (2005)
22	*Cordyline fruticosa*	Burkill (1966); Christman (2000a); Wagner, Herbst and Sohmer (1990); Gilman (1999b); Neal (1965); Radhakrishnan (2003)
23	*Crassula ovata*	Baran (1997–2007); USDA, ARS, National Genetic Resources Program (2007); Mercer (Undated)
24	*Crataegus* species	Radford and Radford (1961); Vickery (1995); Kendall (2002–3); Gu and Spongberg (2003); Phipps, O'Kennon and Lance (2003), Wichman and co-authors (2006); Lassaigne and Blazich (Undated)
25	*Curcuma longa*	Burkill (1966); de Guzman and Siemonsma (1999)
26	*Dendranthema xgrandiflora*	Sze (1963); Tan (1991); Goody (1993); Seaton (1995); Vickery (1995); Gillman (1999c)
27	*Dianthus caryophyllus*	Goody (1993); Seaton (1995); Vickery (1995); Gilman and Delvalle (1999); The United States National Arboretum (2006); South African National Biodiversity Institute (Undated)
28	*Dracaena fragrans*	Wolverton, Douglas, and Bound (1989); Gillman (1999d)
29	*Dracaena sanderana*	–
30	*Epiphyllum oxypetalum*	Burkill (1966); Polunin (1987); Tan (1991); Dave's Garden (2000–2007d–e); Li and Taylor (Undated)
31	*Euphorbia pulcherrima*	Scheper (1998); The University of Illinois Extension (2007)
32	*Ficus benjamina*	Gillman and Watson (1993g); Berg and Corner (2005)
33	*Ficus carica*	Morton (1987); Seaton (1995); Musselman (2000); Kislev, Hartmann and Bar-Yosef (2006)
34	*Ficus microcarpa*	Berg and Corner (2005); Ng and co-authors (2005)
35	*Ficus religiosa*	Gupta (1971); Corner (1983)
36	*Fortunella japonica* and *Fortunella margarita*	Morton (1987); Nguyen and Jansen (1992); Christman (1998)

	Plant Entry	References
37	Four Leaved Clover	Radford and Radford (1961); Wagner, Herbst and Sohmer (1990); Nelson (1991); Vickery (1995); Dave's Garden (2000–2007g, i); BBC News (2004); Bord Bia (Undated); Frame (Undated a, Undated b); North Dakota Department of Agriculture (Undated)
38	*Fuchsia* species and hybrid cultivars	Polomski and Scott (Undated)
39	*Gladiolus* species and hybrids	Folkard (1892); Griffiths (1994); Goody (1993); Missouri Botanical Garden Kemper Center for Home Gardening (2001–2007); Goldblatt (Undated)
40	Heavily Fruiting Plants	–
41	*Hedera helix*	Radford and Radford (1961); Seaton (1995); Vickery (1995); Wichman and co-authors (2006); Institute of Pacific Islands Forestry (2007)
42	*Hordeum vulgare*	Moldenke and Moldenke (1952); Duke (1983); Ceccarelli and Grando (1996); Elbaum (2003)
43	*Hyacinthus orientalis*	Goody (1993); Brunke, Hammerschmidt and Schaus (1994); Seaton (1995); Dave's Garden (2000–2007f); The Columbia Encyclopedia, 6th Edition (2001–05 a, b); Christman (2003a)
44	*Hypericum* species	Radford and Radford (1961); Seaton (1995); Vickery (1995); Christman (2001); Hypericum Depression Trial Study Group (2002); Randløv and co-authors (2006); Li and co-authors (Undated)
45	*Ilex* species and hybrids	Radford and Radford (1961); Hole (1976); Nathan (1988); Wagner, Herbst and Sohmer (1990); Vickery (1995); Hagan (2000); Evans (2000–2003); Kendall (2001–2002); Kluepfel and co-authors (Undated)
46	*Jasminum sambac*	Gimlette (1971); Tan (1991); Burkill (1966); Rahajoe, Kiew and van Valkenburg (1999); Seaton (1995); Christman (2003b)
47	*Juniperus* species	Folkard (1892); Radford and Radford (1961); Rätsch (1992); Seaton (1995); Floridata 2.0 (1996–2007); Kendall (2006); Missouri Botanical Garden Kemper Center for Home Gardening (2001–2007)
48	*Lageneria siceraria*	Widjaja and Reyes (1994); Welman (2005)
49	*Laurus nobilis*	Bulfinch (1855); Goody (1993); Seaton (1995); Vickery (1995); Ipor and Oyen (1999); The Herb Society (2007)
50	*Lilium* species and hybrid cultivars	Radford and Radford (1961); Haw (1986); Seaton (1995); Webber and Cram (2003); Christman (2007); Skinner (Undated)
51	*Mangifera indica*	Gupta (1971); Corner (1988); Frere and Frere (1967); Sukonthasing, Wongrakpanich and Verheij (1992); Gilman and Watson (1994a); Institute of Pacific Islands Forestry, (2007)
52	*Michelia* xalba	Burkill (1966); Corner (1983); Nooteboom 1988); Wee (2003)

	Plant Entry	References
53	*Michelia champaca*	Burkill (1966); Corner (1983); Nooteboom (1988); Tan (1991); Dave's Garden (2000–2007h); Wee (2003)
54	*Musa* species and hybrid cultivars	Gupta (1971); Morton (1987); Wichman and co-authors (2006)
55	*Nelumbo nucifera*	Gupta (1971); Seaton (1995); Barthlott and Neinhaus (1997); Christman (1999a); ASEAN Secretariat (2006); South African National Biodiversity Institute (Undated)
56	*Nyctanthes arbor-tristis*	Burkill (1966); Gupta (1971); Corner (1983); Saxena et al. (1987); Partomihardjo (1991); Simoons (1998); Saxena, Gupta and Lata (2002)
57	*Ocimum tenuiflorum*	Gupta (1971); Stutley (1985); Sunarto (1994); Stearn (1996); Simoons (1998)
58	*Olea europaea*	Seaton (1995); Missouri Botanical Garden Kemper Center for Home Gardening (2001–2007); Reuters (2004); Hemmerly-Brown (2007); Van der Vossen, Mashungwa and Mmolotsi (2007); Columbia Electronic Encyclopedia (Undated)
59	*Oryza sativa*	Burkill (1966); Vergara and de Datta (1996)
60	*Pachira aquatica*	California Rare Fruit Growers, Inc. (1996); Floridata (2000); Liberty Times (2006); Biodiversity International (Undated)
61	*Paeonia* cultivars	Radford and Radford (1961); Stearn and Davis (1984); Goody (1993); Seaton (1995); Vickery (1995); Halevy (1999); Halda & Waddick (2004); The United States National Arboretum (2006)
62	*Phoenix dactylifera*	Morton (1987); Porter (1993); Seaton (1995); Harvard Divinity School (2007); Jewish Virtual Library (2007)
63	*Pinus* species	Mirov and Hasbrouck (1976); Food and Agriculture Organization of the United Nations (1995); Seaton (1995); Vickery (1995); Moussouris and Regato (1999); Earle (2006)
64	*Platycladus orientalis*	Neal (1965); Silba (1996); Fu, Yu and Farjon (1999); Missouri Botanical Garden Kemper Center for Home Gardening (2001–2007); Bielefeldt (2007)
65	Prickly, Spiny or Thorny Plants	Frazer (1935); Bodde (1975); Seaton (1995); Schmidt (1994–2007); Simoons (1998)
66	*Prunus mume* and *Prunus salicina*	Koehn (1952); Sze (1963); Goody (1993); Gilman and Watson (1994b); Missouri Botanical Garden Kemper Center for Home Gardening (2001–2007); Government Information Office, Republic of China (2002); Nanjing International (2002); Gu and Bartholomew (2003); McMahon (2003); Rieger (2006)
67	*Prunus persica*	Wright (1903); Koehn (1952); Antoni (1991); Griffiths (1994); Seaton (1995); Simoons (1998); Missouri Botanical Garden Kemper Center for Home Gardening (2001–2007); Rieger (2006)
68	*Prunus serrulata* and *Prunus xyedoensis*	Griffiths (1994); Seaton (1995); Brand (1997–2001a, b); Missouri Botanical Garden Kemper Center for Home Gardening (2001–2007); McClellan (2005); Lee and co-authors (2007); Mishima (2007); Jefferson (Undated)

	Plant Entry	References
69	*Punica granatum*	Christman (2000b); Morton (1987); Sudiarto and Rifai (1992)
70	*Rosa* species and hybrid cultivars	Radford and Radford (1961); Tergit (1961); Vickery (1995); Gu and Robertson (2003); The United States National Arboretum (2006a); Wichman and co-authors (2006); South African National Biodiversity Institute (Undated)
71	*Rosmarinus officinalis*	Radford and Radford (1961); Baumann (1993); Goody (1993); Seaton (1995); Vickery (1995); Christman (1999b); Guzman (1999)
72	*Saccharum officinarum*	Folkard (1892); Neal (1965); Kuntohartono and Thijsse (1996)
73	*Salix caprea*	Gilman and Watson (1994c); Chao and Gong (1999)
74	*Salix* species and hybrids	Radford and Radford (1961); Gilman and Watson (1994c); Seaton (1995); Vickery (1995); Chao and Gong (1999)
75	*Saraca asoca*	Dutt (1908); Burkill (1966); Gupta (1971); Corner (1988); Maity (1989); Thiselton-Dyer (2004); Andhra Pradesh Medicinal & Aromatic Plants Board (2006); Ali (Undated)
76	*Solanum mammosum*	Neal (1965); Howard (1989)
77	*Thuja occidentalis*	Rätsch (1992); Seaton (1995); Fu, Yu and Farjon (1999); Missouri Botanical Garden Kemper Center for Home Gardening (2001–2007); Iles (2006); Columbia Electronic Encyclopedia (Undated a–b)
78	*Thysanolaena latifolia*	Burkill (1966); Buchar (2002); Duistermaat (2005)
79	*Vicia faba*	Radford and Radford (1961); Jansen (1989); Simoons (1998)
80	*Viscum album*	Bostock and Riley (1855); Baumann (1993); Vickery (1995); Dave's Garden (2000–2007j); Abdulla (Undated)
81	*Vitis vinifera*	Ketsa and Verheij (1992); Corder and co-authors (2006); Wichman and co-authors (2006)
82	*Zamioculcas zamiifolia*	Griffiths (1994); Anderwald (2000–2002)
83	*Zantedeschia aethiopica*	Goody (1993); Seaton (1995); Vickery (1995); Christman (1997c); Charters (2003–2005); Institute of Pacific Islands Forestry (2007)

List of References

Abdulla, P. Undated. *Viscum album* Linn. Flora of Pakistan. http://www.efloras.org/florataxon.aspx?flora_id=5&taxon_id=200006582. (Accessed 21 Oct 2007).

Ali, S.I. Undated. *Saraca asoca* (Roxb.) de Wilde in Blumea. 15: 393. 1967. Zuijd. in Blumea 15: 422. 1967. Caesalpiniaceae. Flora of Pakistan. Vol. 54. http://www.efloras.org/florataxon.aspx?flora_id=5&taxon_id=242425222. (Accessed 5 Nov 2007).

Anderwald, N. 2000–2002. *Zamioculcas* home page. http://members.chello.at/norbert.anderwald/Zamioculcas/index_e.htm. (Accessed 8 Aug 2007).

Andhra Pradesh Medicinal & Aromatic Plants Board. 2006. Ashoka / Ashoka / *Saraca asoca* (Roxb.) de Wilde. Andhra Pradesh Medicinal & Aromatic Plants Board , Andhra Pradesh. http://apmab.ap.nic.in/products.php#. (Accessed 23 Nov 2007).

Antoni, K. 1991. *Momotarō* (The Peach Boy) and the Spirit of Japan: Concerning the function of a fairy tale in Japanese nationalism of the early Sh_wa Age. *Asian Folklore Studies* 50: 155–188.

ASEAN Secretariat. 2006. ASEAN national flowers. http://www.aseansec.org/18203.htm. (Accessed 25 Jul 2007).

Ashari, S. 1992. *Citrus reticulata* Blanco. In: Verjeij, E.W.M. and R.E. Coronel (Editors). Plant resources of South-East Asia. No. 2. Edible fruits and nuts. PROSEA, Bogor. Pages 135–138.

Backer, C.A. 1936. *Verklarend woordenboek der wetenschappelijke namen van de in Nederland en Nederlandsch-Indië in het wild groeiende en in tuinen en parken gekweekte varens en hoogere planten*. P. Noordhoff N.V., Groningen, Noordhoff-Kolff Batavia (C.) and Visser & Co., Batavia.

Baran, R.J. 1999–2007. *Portulacaria afra*, the Elephant's Food or Spekboom: a monograph which contains some of the areas of both knowledge and ignorance pertaining to this plant. http://www.phoenixbonsai.com/Portulacaria.html. (Accessed 8 Aug 2007).

Barthlott, W. and C. Neinhaus. 1997. Purity of the sacred lotus, or escape from contamination in biological surfaces. *Planta (Heidelberg)* 202: 1–8.

Bates, R. 2003. Caring for your cut Christmas tree. Department of Horticulture Fact Sheet, Pennsylvania State University, Pennsylvania. http://consumerhorticulture.psu.edu/pdfs/christmas_tree.pdf. (Accessed 27 November 2007).

Baumann, H. 1993. The Greek plant world in myth, art and literature. Translated and augmented by W.T. Stearn and E.R. Stearn. Timber Press, Portland, Oregon.

BBC News. 2005. Japan bean plants sprout messages. British Broadcasting Corporation, London. http://news.bbc.co.uk/2/hi/asia-pacific/4209571.stm. (Updated 26 January 2005). (Accessed 10 November 2007).

BBC News. 2004. The truth behind the shamrock. British Broadcasting Corporation, London. http://news.bbc.co.uk/1/hi/uk/3519116.stm. (Accessed 10 October 2007).

Berg, C.C. and E.J.H. Corner. 2005. Moraceae – *Ficus*. Flora Malesiana Series I, 17: 1–730.

Bielefeldt, C. 2007. Notes on the translation of Dogen. 1242. Treasury of the Eye of the True Dharma. Number 40. The cypress tree. Sotoshu Shumucho, Japan. http://www.stanford.edu/group/scbs/sztp3/translations/shobogenzo/translations/hakujushi/pdf/notes.pdf. (Accessed 21 Nov 2007).

Biodiversity International. Undated. Bombacaceae *Pachira aquatica*. http://www.bioversityinternational.org/Information_Sources/Species_Databases/New_World_Fruits_Database/qryall3.asp?intIDSpecies=285. (Accessed 21 Aug 2007).

Bodde, D. 1975. Festivals in classical China: New Year and other annual observances during the Han dynasty, 206 B.C.–A.D. 220. Princeton University Press, Princeton.

Bone, S. Undated. Artificial and real Christmas trees compared. Christmas Tree Council of Nova Scotia, Nova Scotia. http://www.ctcns.com/fake-real.pdf. (Accessed 5 Dec 2007).

Boo, C.M., Kartini Omar-Hor and C.L. Au-Yang. 2006. 1001 garden plants in Singapore. 2nd edition. National Parks Board, Singapore.

Bosch, C.H. 2004. *Canavalia gladiata* (Jacq.) DC. [Internet] Record from Protabase. Grubben, G.J.H. & Denton, O.A. (eds). PROTA (Plant Resources of Tropical Africa / Ressources végétales de l'Afrique tropicale), Wageningen, Netherlands. http://database.prota.org/search.htm. (Accessed 12 November 2007).

Bostock J. and H.T. Riley (Editors). 1855. Pliny the Elder, *The Natural History*. Taylor and Francis, London. http://www.perseus.tufts.edu/cgi-bin/ptext?lookup=Plin.+Nat.+toc. (Accessed19 Oct 2007).

Bord Bia. Undated. Shamrock. http://www.bordbia.ie/go/Consumers/Plants_Flowers/shamrock.htm. (Irish Food Board). (Accessed 20 Nov 2007).

Brand, M.H. 1997–2001a. *Prunus serrulata.* UConn Plant Database of Trees, Shrubs and Vines. http://www.hort.uconn.edu/plants/p/pruerr/pruerr1.html. (Accessed 28 November 2007).

Brand, M.H. 1997–2001b. *Prunus xyedoensis*. UConn Plant Database of Trees, Shrubs and Vines. http://www.hort.uconn.edu/Plants/p/pruyed/pruyed1.html. (Accessed 29 November 2007).

Brunke, E.-J., F.-J. Hammerschmidt and G. Schmaus. 1994. Headspace analysis of hyacinth flowers. *Flavour and Fragrance Journal* 9: 59–69.

Buchar, S.K. 2002. An ecophysiological study of Thysanolaena maxima (broom grass): a multi-purpose grass of high fodder value. ENVIS Centre@GBPIHED. http://gbpihed.nic.in/envis/thesis/thesis_bhuchar.html. (Accessed 11 Aug 2007).

Bulfinch, T. 1855. *The Age of Fable*. http://www.sacred-texts.com/cla/bulf/. (Accessed 8 Oct 2007).

Burkill, I.H. 1966. A dictionary of the economic products of the Malay Peninsula. 2nd edition. Volumes I and II. Ministry of Agriculture and Co-operatives on behalf of the governments of the Malaysia and Singapore, Kuala Lumpur.

California Rare Fruit Growers, Inc. 1996. Malabar chestnut. http://www.crfg.org/pubs/ff/malabar.html. (Accessed 21 Aug 2007).

Ceccarelli, S. and S. Grando. 1996. *Hordeum vulgare* L. In: Grubben, G.J.H. and S. Partohardjono (Editors). Plant Resources of South-East Asia. No. 10. Cereals. PROSEA, Bogor. Pages 99–102.

Chao, N. and G.T. Gong. 1999. 3. *Salix* Linnaeus, Sp. Pl. 2: 1015. 1753. Flora of China 4: 162–274. http://flora.huh.harvard.edu/china/PDF/PDF04/salix.pdf. (Accessed 13 Nov 2007).

Charters, M.L. 2003–2005. California Plant Names: Latin and Greek Meanings and Derivations: A Dictionary of Botanical Etymology. http://www.calflora.net/botanicalnames/. (Updated 1 Jul 2007). (Accessed 22 Aug 2007).

Chastagner, G.A. and D.M. Benson. 2000. The Christmas Tree: Traditions, Production, and Diseases. *Plant Health Progress* 13 October 2000. http://www.plantmanagementnetwork.org/pub/php/review/1225tree/. (Accessed 26 November 2007).

Chastagner, G. and E. Hinesley. Undated. How to care for your farm-grown fresh Christmas tree. http://www.christmastree.org/care.cfm. (Accessed 27 November 2007).

Chiceley Plowden, C. 1968. A manual of plant names. George Allen & Unwin, Ltd., London.

Christman, S. 1997a. *Camellia japonica*. Floridata. (Updated 3 Dec 2003). http://www.floridata.com/ref/C/cam_jap.cfm. (Accessed 30 Jul 2007).

Christman, S. 1997b. *Hedera helix*. Floridata. (Updated 6 May 2006). http://www.floridata.com/ref/h/hedera_h.cfm. (Accessed 26 Oct 2007).

Christman, S. 1997c. *Zantedeschia aeithiopica*. Floridata. (Updated 22 Jun 2006). http://www.floridata.com/ref/z/zant_aet.cfm. (Accessed 22 Aug 2007).

Christman, S. 1998. *Fortunella* spp. Floridata. (Updated 26 Jan 2005). http://www.floridata.com/ref/F/fort_mar.cfm. (Accessed 21 Aug 2007).

Christman, S. 1999a. *Nelumbo lutea*. Floridata. (Updated 26 Feb 2005). http://www.floridata.com/ref/n/nelu_lut.cfm. (Accessed 4 Aug 2007).

Christman, S. 1999b. *Rosmarinus officinalis*. Floridata. (Updated 17 Aug 2004). http://www.floridata.com/ref/r/rose_off.cfm. (Accessed 15 Nov 2007).

Christman, S. 2000a. *Cordyline fruticosa*. Floridata. (Updated 6 May 2006). http://www.floridata.com/ref/c/cord_fru.cfm. (Accessed 23 Jul 2007).

Christman, S. 2000b. *Punica granatum*. Floridata. (Updated 20 Sep 2003). http://www.floridata.com/ref/p/puni_gra.cfm. (Accessed 13 Jan 2008).

Christman, S. 2001. *Hypericum perforatum*. Floridata. (Updated 16 Nov 2007). http://www.floridata.com/ref/H/hype_per.cfm. (Accessed 16 Jul 2007).

Christman, S. 2003a. *Hyacinthus orientalis*. Floridata. http://www.floridata.com/ref/h/hyac_ori.cfm (Updated 26 Jan 2004). (Accessed 9 Oct 2007).

Christman, S. 2003b. *Jasminum sambac.* Floridata. http://www.floridata.com/ref/j/jasm_sam.cfm. (Accessed 21 Oct 2007).
Christman, S. 2007. *Lilium* hybrids. Floridata. http://www.floridata.com/ref/l/lili_hyb.cfm. (Accessed 26 Oct 2007).
Chongqing Municipal Government. 2007. City flower and city tree. http://english.cq.gov.cn/ChongqingGuide/MountainCity/1914.htm. (Accessed 30 Jul 2007).
Chua, K.S., Soong, B.C., Tan, H.T.W. 1996. The Bamboos of Singapore. International Plant Genetic Resources Institute, Singapore.
Ciesla, W.M. 2002. Non Wood Forest Products from Temperate Broad-Leaves Trees. *Non-Wood Forest Products* 15. Food and Agriculture Organization of The United Nations, Rome.
Columbia Electronic Encyclopedia. Undated a. Cartier, Jacques. Columbia Electronic Encyclopedia. Columbia University Press. http://www.reference.com/browse/columbia/CartierJ (Accessed 21 Nov 2007).
Columbia Electronic Encyclopedia. Undated b. Francis I. Columbia Electronic Encyclopedia. Columbia University Press. http://www.reference.com/browse/columbia/Francis1Fr. (Accessed 21 Nov 2007).
Columbia Electronic Encyclopedia. Undated c. Olive. Columbia Electronic Encyclopedia. Columbia University Press, Columbia. http://www.reference.com/browse/columbia/olive. (Accessed 21 Nov 2007).
Corder, R., W. Mullen, N.Q. Khan, S.C. Marks, E.G. Wood, M.J. Carrier and A. Crozier. 2006. Oenology: red wine procyanidins and vascular health. Nature 444: 566.
Corner, E.J.H. 1983. Wayside trees of Malaya. 3rd edition. Volumes 1 and 2. Malaya Nature Society, Kuala Lumpur.
Darlington & Stockton Times. 2006. Why picking a milkmaid or lady's smock is just asking for trouble.
Darlington & Stockton Times. 23 Jun 2006. http://www.darlingtonandstocktontimes.co.uk/search/display.var.806082.0.why_picking_a_milkmaid_or_ladys_smock_is_just_asking_for_trouble.php. (Accessed 22 October 2007).
Dave's Garden. 2000–2007a. Guides for information. *Cardamine pratensis.* http://davesgarden.com/guides/pf/go/1542/. (Accessed 22 Oct 2007).
Dave's Garden. 2000–2007b. Guides for information. *Cercis siliquastrum.* http://davesgarden.com/guides/pf/go/53634/. (Accessed 5 Nov 2007).
Dave's Garden. 2000–2007c. Guides for information. *Conium maculatum.* http://davesgarden.com/guides/pf/go/2794/. (Accessed 8 Oct 2007).
Dave's Garden. 2000–2007d. Guides for information. *Epiphyllum oxypetalum.* http://davesgarden.com/guides/pf/go/2443/. (Accessed 15 Nov 2007).
Dave's Garden. 2000–2007e. Guides for information. *Epiphyllum oxypetalum* var. *purpusii.* http://davesgarden.com/guides/pf/go/130583/. (Accessed 15 Nov 2007).
Dave's Garden. 2000–2007f. Guides for information. *Hyacinthus orientalis* var. *albulus.* http://davesgarden.com/guides/pf/go/764/. (Accessed 8 Oct 2007).
Dave's Garden. 2000–2007g. Guides for information. *Medicago lupulin*a. http://davesgarden.com/guides/pf/go/764/. (Accessed 20 Nov 2007).
Dave's Garden. 2000–2007h. Guides for information. *Michelia champaca.* http://davesgarden.com/guides/pf/go/53529/. (Accessed 24 Aug 2007).
Dave's Garden. 2000–2007i. Guides for information. *Oxalis acetosella.* http://davesgarden.com/guides/pf/go/2043/. (Accessed 20 Nov 2007).
Dave's Garden. 2000–2007j. Guides for information. *Viscum album.* http://davesgarden.com/guides/pf/go/2787/. (Accessed 21 Oct 2007).
Debevec, M. 2002. *Castanospermum australe.* Australian National Botanic Garden, Canberra. http://www.anbg.gov.au/gnp/interns-2002/castanospermum-australe.html. (Updated 16 Dec 2003). (Accessed 14 Oct 2007).
Duistermaat, H. 2005. Field guide to the grasses of Singapore. Supplement to The Gardens' Bulletin Singapore 57: 1–177.
Duke, J.A. 1983. *Ananas comosus* (L.) Merr. Handbook of Energy Crops. Unpublished. http://www.hort.purdue.edu/newcrop/duke_energy/Ananas_comosus.html. (Accessed 29 Jul 2007).
Duke, J.A. 1983. *Hordeum vulgare* L. Handbook of Energy Crops. Unpublished. http://www.hort.purdue.edu/newcrop/nexus/Hordeum_vulgare_nex.html. (Accessed 16 Oct 2007).
Dutt, M.N. 1908. *Garuda Purana.* Society for the Resurrection of Indian Literature, Calcutta.

Earle, C.J. (Editor). 2006. *Pinus* description. The Gymnosperm Database. (Updated 28 May 2007) http://www.conifers.org/pi/pin/index.htm. (Accessed 31 October 2007).

Earle, C.J. (Editor). 2004a. *Abies balsamea* description. Gymnosperm Database. http://www.conifers.org/pi/ab/balsamea.htm (Accessed 27 November 2007).

Earle, C.J. (Editor). 2004b. *Abies fraseri* description. Gymnosperm Database. http://www.conifers.org/pi/ab/fraseri.htm (Accessed 27 November 2007).

Earle, C.J. (Editor). 2004c. *Abies nordmanniana* description. Gymnosperm Database. http://www.conifers.org/pi/ab/nordmanniana.htm (Accessed 27 November 2007).

Earle, C.J. (Editor). 2004d. *Picea pungens* description. Gymnosperm Database. http://www.conifers.org/pi/pic/pungens.htm (Accessed 27 November 2007).

Earle, C.J. (Editor). 2005. *Abies procera* description. Gymnosperm Database. http://www.conifers.org/pi/ab/procera.htm (Accessed 27 November 2007).

Earle, C.J. (Editor). 2006a. *Abies alba* description. Gymnosperm Database. http://www.conifers.org/pi/ab/alba.htm (Accessed 27 November 2007).

Earle, C.J. (Editor). 2006b. *Picea abies* description. Gymnosperm Database. http://www.conifers.org/pi/pic/abies.htm (Accessed 27 November 2007).

Earle, C.J. (Editor). 2006c. *Pseudotsuga menziesii* description. Gymnosperm Database. http://www.conifers.org/pi/ps/menziesii.htm (Accessed 27 November 2007).

Earle, C.J. (Editor). 2007. *Pinus sylvestris* description. Gymnosperm Database. http://www.conifers.org/pi/pin/sylvestris.htm (Accessed 27 November 2007).

Elbaum, L. 2003. Plants of the Bible. http://www.geocities.com/jelbaum/plants2002.html#seven. (Updated 8 Apr 2003). (Accessed 19 Oct 2007).

Evans, E. 2000–2003. *Ilex aquifolium*. NC State University. http://www.ces.ncsu.edu/depts/hort/consumer/factsheets/trees-new/ilex_aquifolium.html. (Accessed on 12 Nov 2007).

Fagg, M. 2002. Australian Christmas plants. http://www.anbg.gov.au/christmas/index.html. (Accessed 26 November 2007).

Faucon, P. 1998–2005. Desert rose. http://www.desert-tropicals.com/Plants/Apocynaceae/Adenium_obesum.html. (Accessed 3 Aug 2007).

Floridata 2.0. 2000. *Pachira aquatica*. Floridata.com LC. http://www.floridata.com/ref/p/pach_aqu.cfm. (Accessed 21 Aug 2007).

Floridata 2.0. 1996–2007. Plant encyclopedia profiles. Floridata.com LC. http://www.floridata.com/index.cfm. (Accessed 19 Nov 2007).

Folkard, R. 1892. Plant lore, legends and lyrics: embracing the myths, traditions, superstitions, and folk-lore of the plant kingdom. Sampson, Low, Marston, London. http://dlxs2.library.cornell.edu/cgi/t/text/text-idx?c=cdl;cc=cdl;view=toc;subview=short;idno=cdl436. (Accessed 20 Nov 2007).

Food and Agriculture Organization of the United Nations. 1995. Non-wood forest products from conifers. Non-Wood Forest Products 12. http://www.fao.org/docrep/X0453E/X0453E00.htm. (Accessed 31 October 2007).

Frame, J. Undated. Grassland Species Profiles. *Trifolium dubium* Sibth. CIAT/FAO Collaboration on Tropical Forages. http://www.fao.org/AG/AGP/AGPC/doc/Gbase/DATA/Pf000499.htm. (Accessed 20 Nov 2007).

Frame, J. Undated. Grassland Species Profiles. *Trifolium repens* L. CIAT/FAO Collaboration on Tropical Forages. http://www.fao.org/AG/AGP/AGPC/doc/Gbase/data/pf000350.htm. (Accessed 20 Nov 2007).

Fraser, A. 2002. Judas Tree – *Cercis siliquastrum*. http://www.the-tree.org.uk/BritishTrees/TreeGallery/judasc.htm. (Accessed 22 October 2007).

Frazer, J.G. 1935. The golden bough, a study in magic and religion. 3rd Edition. 12 Volumes. MacMillan, New York.

Frere, M. and B. Frere. 1967. Old Deccan Days. Dover Publications, New York.

Fu, L., Y. Yu and A. Farjon. 1999. Cupressaceae. Flora of China 4: 62–77. http://flora.huh.harvard.edu/china/mss/volume04/CUPRESSACEAE.published.pdf. (Accessed 21 Nov 2007).

Furlow, J.J. Undated. 31. Betulaceae Gray. Flora of North America 3: http://www.efloras.org/florataxon.aspx?flora_id=1&taxon_id=10101. (Accessed 10 Nov 2007).

Gilman, E.F. 1999a. Fact Sheet FPS-97. *Camellia japonica*. University of Florida, Institute of Food and Agricultural Services, Cooperative Extension Service. http://hort.ifas.ufl.edu/shrubs/CAMJAPA.PDF. (Accessed 30 Jul 2007).

Gilman, E.F. 1999b. Fact Sheet FPS-141. *Cordyline terminalis*. University of Florida, Institute of Food and Agricultural Services, Cooperative Extension Service. http://hort.ifas.ufl.edu/shrubs/CORTERA.PDF. (Accessed 23 Jul 2007).

Gilman, E.F. 1999c. Fact Sheet FPS-171. *Dendranthema* x*grandiflora*. University of Florida, Institute of Food and Agricultural Services, Cooperative Extension Service. http://hort.ifas.ufl.edu/shrubs/DENGRAA.PDF. (Accessed 9 Aug 2007).

Gilman, E.F. 1999d. Fact Sheet FPS-184. *Dracaena fragrans* 'Massangeana'. University of Florida, Institute of Food and Agricultural Services, Cooperative Extension Service. http://hort.ifas.ufl.edu/shrubs/DRAFRAA.PDF. (Accessed 17 Aug 2007).

Gilman, E.F. and T. Delvalle. 1999. Fact Sheet FPS-173. *Dianthus* x 'Parfait Series'. University of Florida, Institute of Food and Agricultural Services, Cooperative Extension Service. http://hort.ifas.ufl.edu/shrubs/DIASPPA.PDF. (Accessed 25 Jul 2007).

Gilman, E.F. and D.G. Watson. 1993a. Fact Sheet ST-94. *Betula nigra*: River Birch. University of Florida, Institute of Food and Agricultural Sciences. (November 1993). http://hort.ifas.ufl.edu/trees/BETNIGA.pdf. (Accessed 10 Nov 2007).

Gilman, E.F. and D.G. Watson. 1993b. Fact Sheet ST-96. *Betula papyrifera*: Paper Birch. University of Florida, Institute of Food and Agricultural Sciences. (November 1993). http://hort.ifas.ufl.edu/trees/BETPAPA.pdf. (Accessed 10 Nov 2007).

Gilman, E.F. and D.G. Watson. 1993c. Fact Sheet ST-97. *Betula pendula*: European Birch. University of Florida, Institute of Food and Agricultural Sciences. (November 1993). http://hort.ifas.ufl.edu/trees/BETPENA.pdf. (Accessed 10 Nov 2007).

Gilman, E.F. and D.G. Watson. 1993d. Fact Sheet ST-98. *Betula populifolia*: Gray Birch. University of Florida, Institute of Food and Agricultural Sciences. (November 1993). http://hort.ifas.ufl.edu/trees/BETPOPA.pdf. (Accessed 10 Nov 2007).

Gilman, E.F. and D.G. Watson. 1993e. Fact Sheet ST-169. *Citrus* spp.: citrus. University of Florida, Institute of Food and Agricultural Sciences. (November 1993). http://hort.ifas.ufl.edu/trees/CITSPPA.pdf. (Accessed 21 Aug 2007).

Gilman, E.F. and D.G. Watson. 1993f. Fact Sheet ST-177. *Cocos nucifera*: coconut palm. University of Florida, Institute of Food and Agricultural Sciences. (November 1993). http://hort.ifas.ufl.edu/trees/COCNUCA.pdf. (Accessed 7 Aug 2007).

Gilman, E.F. and D.G. Watson. 1993g. Fact Sheet ST-251. *Ficus benjamina*: weeping fig. University of Florida, Institute of Food and Agricultural Sciences. (November 1993). http://hort.ifas.ufl.edu/trees/FICBENA.pdf. (Accessed 14 Jul 2007).

Gilman, E.F. and D.G. Watson. 1994a. Fact Sheet ST-404. *Mangifera indica*: mango. University of Florida, Institute of Food and Agricultural Sciences. (October 1994). http://hort.ifas.ufl.edu/trees/MANINDA.pdf. (Accessed 2 Nov 2007).

Gilman, E.F. and D.G. Watson. 1994b. Fact Sheet ST-512. *Prunus mume*: Japanese apricot. University of Florida, Institute of Food and Agricultural Sciences. (October 1994). http://hort.ifas.ufl.edu/trees/MANINDA.pdf. (Accessed 5 Dec 2007).

Gilman, E.F. and D.G. Watson. 1994c. Fact Sheet ST-576. *Salix* spp.: weeping willow. University of Florida, Institute of Food and Agricultural Sciences. (October 1994). http://hort.ifas.ufl.edu/trees/SALSPPA.pdf. (Accessed 13 Nov 2007).

Gimlette, J.D. 1971. Malay poisons and charm cures. Oxford University Press, London.

Goldblatt, P. Undated. 16. *Gladiolus* Linnaeus, Sp. Pl. 1: 36. 1753; Gen. Pl. ed. 5, 23. 1754. Flora of North America 26: 407. http://www.efloras.org/florataxon.aspx?flora_id=1&taxon_id=113591. (Accessed 20 Nov 2007).

Goody, J. 1993. The culture of flowers. Cambridge University Press, Cambridge.

Government Information Office, Republic of China. 2002. National Flower. Government Information Office, Republic of China. http://www.gio.gov.tw/taiwan-website/5-gp/yearbook/2002/national_flower.htm. (Accessed 4 Dec 2007).

Griffiths, M. 1992. Index of garden plants. Timber Press, Portland, Oregon.

Gu, C and B. Bartholomew. 2003. 50. *Prunus* Linnaeus, Sp. Pl. 1: 473. 1753. Flora of China 9: 401–403. http://flora.huh.harvard.edu/china/PDF/PDF09/Prunus.PDF. (Accessed 5 Dec 2007).

Gu, C. and K.R. Robertson. 2003. 41. *Rosa* Linnaeus, Sp. Pl. 1: 491. 1753. Flora of China 9: 339.381.

Gu, C. and S.A. Spongberg. 2003. 12. *Crataegus* Linnaeus, Sp. Pl. 1: 475. 1753. Flora of China 9: 111–117.

Gupta, S.M. 1971. Plant myths and traditions in India. E.J. Brill, Leiden.
Guzman, C.C. de. 1999. *Rosmarinus officinalis* L. In: de Guzman, C.C. and J.S. Siemonsma (Editors). Plant Resources of South-East Asia. No. 13. Spices. PROSEA, Bogor. Pages 194–197.
Hagan, A. 2000. ANR-1087. Common diseases of holly and their control. Alabama Cooperative Extension System. Alabama A & M and Auburn Universities. http://www.aces.edu/pubs/docs/A/ANR-1087/ANR-1087.pdf. (Accessed 12 Nov 2007).
Halda, J.J. and J.W. Waddick. 2004. The genus *Peonia*. Timber Press, Portland.
Halevy, A.H. 1999. Ornamentals: Where diversity is king–the Israeli experience. In: J. Janick (Editor). Perspectives on new crops and new uses. ASHS Press, Alexandria, VA. Pages 404–406.
Harvard Divinity School. 2007. Multifaith calendar glossary. President and Fellows of Harvard College. http://www.hds.harvard.edu/spiritual/calendar_glossary.html. (Accessed 19 Nov 2007).
Haw, S.G. 1986. The lilies of China: The genera *Lilium*, *Cardiocrinum*, *Nomocharis* and *Notholirion*. Timber Press, Portland, Oregon, USA.
Healey Dell Nature Reserve. 2006. Field guides to Healey Dell – Plants in early summer: lady's smock. http://www.healeydell.org.uk/files/Ladys_Smock.pdf. (Accessed 22 October 2007).
Hemmerly-Brown, A. 2007. 164th soldiers plant olive trees as symbol of hardiness and peace. AUSA (Association of the United States Army) Greater Los Angeles Chapter Newsletter 40:1, 7. http://www.ausa.org/chapweb/glc/glac/pdfs/spring_07.pdf. (Accessed 20 Nov 2007).
Hole, C. 1976. British folk customs. London.
Holttum, R.E. 1968. A revised flora of Malaya :an illustrated systematic account of the Malayan flora, including commonly cultirated plants.Vol. 2.Ferns of Malaya. 2nd Edition. Government Printing Office, Singapore.
Hora, B. 1981. The Oxford Encyclopedia of trees of the World. Oxford University Press, Oxford.
Howard, Richard A. 1989. Flora of the Lesser Antilles: Leeward and Windward Islands. Vol. 6 Dicotyledoneae. Part 3. Arnold Arboretum, Harvard University, Massachusetts.
Hypericum Depression Trial Study Group 2002. Effect of *Hypericum perforatum* (St. John's wort) in major depressive disorder. The Journal of the American Medical Association 287: 1807–1814.
Iles, J. 2006. *Thuja occidentalis* 'Hetz Wintergreen' (Eastern Arborvitae). Bicklehaupt Arboretum, Clinton, Iowa. http://www.bickarb.org/plantMonth/december_2006.html. (Accessed 21 Nov 2007).
Institute of Pacific Islands Forestry. 2007. Pacific Islands Ecosystems at Risk (PIER) Website version 5.1. http://www.hear.org/Pier/index.html. (Accessed 26 Oct 2007).
Ipor, I.B. and L.P.A. Oyen. 1999. *Laurus nobilis* L. In: de Guzman, C.C. and J.S. Siemonsma (Editors). Plant Resources of South-East Asia. No. 13. Spices. PROSEA, Bogor. Pages 134–137.
Jaeger, E.C. A source-book of biological names and terms. 2nd Edition. Charles C. Thomas, Publisher, Springfield, Illinois.
Jansen, P.C.M. 1989. *Vicia faba* L. In: van der Maesen, L.J.G. and S. Sommatmadja (Editors). Plant Resources of South-East Asia. No. 1. Pulses. PROSEA, Bogor, Indonesia. Pages 64–66.
Jefferson, R.M. Undated. The history of the cherry blossom trees in Potomac Park. http://www.us.emb-japan.go.jp/jicc/sakuralecture.htm. (Accessed 29 Nov 2007).
Jewish Virtual Library. 2007. Sukkot. The American-Israeli Cooperative Enterprise. http://www.jewishvirtuallibrary.org/jsource/Judaism/holiday5.html. (Accessed 19 Nov 2007).
Jones, D.T. 1992. *Citrus medica* L. In: Verheij, E.W.M. and R.E. Coronel (Editors). Plant resources of South-East Asia. No. 2. Edible fruits and nuts. PROSEA, Bogor, Indonesia. Pages 131–133.
Keen, J. 2007. All is bright (and less costly) with LED lights ; Many cities using energy-saving bulbs in holiday decor. USA Today (21 November 2007).
Kendall, P. 2001. Mythology and folklore of the juniper. http://www.treesforlife.org.uk/forest/mythfolk/juniper.html. (Last updated 5 Dec 2006). (Accessed 4 Aug 2007).
Kendall, P. 2001–2. Mythology and folklore of the holly. http://www.treesforlife.org.uk/forest/mythfolk/holly.html. (Last updated 21 Nov 2006). (Accessed 12 Nov 2007).
Kendall, P. 2002–3. Mythology and folklore of the hawthorn. http://www.treesforlife.org.uk/forest/mythfolk/hawthorn.html. (Last updated 21 Nov 2006). (Accessed 4 Aug 2007).
Kendall, P. 2006. Mythology and folklore of the birch. http://www.treesforlife.org.uk/forest/mythfolk/birch.html. (Accessed 29 Oct 2007).
Ketsa, S. and E.W.M. Verheij. 1992. *Vitis vinifera* L. In: Verjeij, E.W.M. and R.E. Coronel (Editors). Plant resources of South-East Asia. No. 2. Edible fruits and nuts. PROSEA, Bogor. Pages 304–310.
Kislev, M.E., A. Hartmann and O. Bar-Yosef. 2006. Eary domestic fig in the Jordan Valley. Science 312: 1372–1374.

Kluepfel, M., J. McLeod Scott, J.B. Blake, C.S. Gorsuch. Undated. HGIC-2055. Holly diseases & insect pests. Revised by J. Williamson. The Clemson University Cooperative Extension Service. http://hgic.clemson.edu/pdf/hgic2055.pdf. (Accessed 12 Nov 2007).

Koehn, A. 1952. Chinese flower symbolism. *Monumenta Nipponica* 8: 121–146.

Kooi, G. 1994. *Canavalia gladiata* (Jacq.) DC. In: Siemonsma, J.S. and K. Piluek (Editors). Plant resources of South-East Asia. No. 8. Vegetables. PROSEA, Bogor, Indonesia. Pages 134–136.

Kuntohartono, T. and J.P. Thijsse. 1996. *Saccharum officinarum* L. In: Flach, M. and F. Rumawas (Editors). Plant resources of South-East Asia. No. 9. Plants yielding non-seed carbohydrates. PROSEA, Bogor, Indonesia. Pages 143–148.

Lai, J. (Director). 2003. Gods must be hungry. Oak 3 Films/Economic Development Board, Singapore.

Lasseigne, F.T. and F.A. Blazich. *Crataegus* L. In: Nisley, R.G. (Managing Editor). USDA woody plant seed manual. http://www.nsl.fs.fed.us/wpsm/Crataegus.pdf. (Accessed 24 Oct 2007).

Lee, B.-B., M.-R. Cha, S.-Y. Kim, E. Park, H.-R. Park and S.-C. Lee. 2007. Antioxidative and anticancer activity of extracts of cherry (*Prunus serrulata* var. *spontanea*) blossoms. Plant Foods for Human Nutrition 62: 79–84.

Li, Z. and N.P. Taylor. Undated. Cactaceae. Flora of China 13: 209–212. http://flora.huh.harvard.edu/china/mss/volume13/Cactaceae.pdf. (Accessed 15 Nov 2007).

Li, X., J. Li, N.K.B. Robson and P.F. Stevens. (Undated). Clusiaceae. Flora of China 13: 1–47. http://flora.huh.harvard.edu/china/mss/volume13/Clusiaceae.pdf. (Accessed 16 Nov 2007).

Liberty Times. 2006. http://english.www.gov.tw/TaiwanHeadlines/index.jsp?recordid=92685. (23 Mar 2006). (Accessed 21 Aug 2007).

Mabberley, D.J. 1997. The plant book: a portable dictionary of the vascular plants utilising Kubitzki's The families and genera of vascular plants (1990), Cronquist's An integrated system of classification of flowering plants (1981), and current botanical literature arranged largely on the principles of editions 1-6 (1896/97–1931) of Willis's A dictionary of the flowering plants and ferns. 2nd edition. Cambridge University Press, New York.

Maheshwari, P. and U. Singh. 1965. Dictionary of economic plants in India. Indian Council of Agricultural Research, New Delhi.

Maity P.K. 1989. Human fertility cults and rituals of Bengal: A comparative study. Ahhinav Publications, New Delhi.

Matsuyama City Office. Undated. A guided information of Matsuyama City. http://www.city.matsuyama.ehime.jp/eng/frame.html. (Accessed 30 Jul 2007).

McClellan, A. 2005. The cherry blossom festival: a sakura celebration. Bunker Hill Publishing, Boston. McMahon, C.G. 2003. The sign system in Chinese landscape paintings. The Journal of Aesthetic Education 37: 64–76.

Mercer, J. Undated. *Crassula ovata*. http://www.bcss-liverpool.pwp.blueyonder.co.uk/Growing_Crassula_ovata.htm. (Accessed 8 Aug 2007).

Mirov, N.T. and Hasbrouck, J. 1976. The story of pines. Indiana University Press, Indiana.

Mishima, S. 2007. Japanese cherry blossom flowering forecast 2007. http://gojapan.about.com/library/special/blkaikayosou2007.htm. (Accessed 28 November 2007).

Missouri Botanical Garden Kemper Center for Home Gardening. 2001–2007. Missouri Botanical Garden. http://www.mobot.org/gardeninghelp/plantfinder/serviceplantfinder.shtml. (Accessed 19 Nov 2007).

Moldenke, H.N. and A.L. Moldenke. 1952. Plants of the Bible. The Ronald Press Company, New York.

Morton, J.F. 1987. Fruits of warm climates. Julia F. Morton, Miami, Florida.

Moussouris, Y. and Regato, P. 1999. Forest Harvest: An Overview of Non Timber Forest Products in the Mediterranean Region. http://www.fao.org/docrep/x5593e/x5593e03.htm. (Accessed 31 October 2007).

Musselman, L.J. 2000. Some plants of the Qu'ran and the Bible: overview and recent research. http://www.odu.edu/~lmusselm/essays/plantsofthebibleandkoranenglish.html. (Accessed 11 Nov 2007).

Nanjing International. 2002. The city flower of Nanjing. The Official Nanjing City Website. www.nj.gov.cn. http://www.nanjing.gov.cn/pub/english/today/gynj/Symbols/200705/t20070521_213199.htm. (Accessed 4 Dec 2007).

Nathan, V.K. 1988. Red-berried hollies. The Virginia Gardener Newsletter 7. http://www.ext.vt.edu/departments/envirohort/factsheets2/landsnurs/dec93pr1.html. (Accessed 12 Nov 2007).

National Fire Protection Association. 2007. Christmas tree fires. National Fire Protection Association, Massachusetts. http://www.nfpa.org/categoryList.asp?categoryID=296&URL=Research%20&%20Reports/Fact%20sheets/Seasonal%20safety/Christmas%20tree%20fires (Updated Nov 2007). (Accessed 28 Nov 2007).

National Research Council. 1992. Neem: A tree for solving global problems. National Academy Press, Washington, D.C. http://www.nap.edu/books/0309046866/html/. (Accessed 25 Jul 2007).

Neal, M.C. 1965. In gardens of Hawaii. Bishop Museum Press, Honolulu.

Nelson, E.C. 1991. Shamrock: botany and history of an Irish myth. Boethius Press, Aberystwyth and Kilkenny.

Nix, S. Undated. Top ten Christmas tree species sold in North America. http://forestry.about.com/cs/christmastrees1/a/top10_xmastree.htm (Accessed 25 November 2007).

Ng, A.B.C., A. Ng, B. Lee, A.L. Chuah, S.G. Goh, J.T.K. Lai, G.C. Tan and V. D'Rozario. 2005. The fabulous figs of Singapore. Singapore Science Centre, Singapore.

Nguyen, T.H. and P.C.M. Jansen. 1992. *Fortunella* Swingle. In: Verheij, E.W.M. and R.E. Coronel (Editors). Plant resources of South-East Asia. No. 2. Edible fruits and nuts. PROSEA, Bogor, Indonesia. Pages 169–171.

North Dakota Department of Agriculture. Undated. Invasive species web-based manual. Black medic (*Medicago lupulina*). North Dakota Department of Agriculture. http://www.agdepartment.com/noxiousweeds/pdf/Blackmedic.pdf. (Accessed 20 Nov 2007).

Nooteboom, H.P. 1988. Magnoliaceae. Flora Malesiana, Series I, 10: 561–605.

Pan, Z. and M.F. Watson. 2005. 31. *Conium* Linnaeus, Sp. Pl. 1: 243. 1753. Flora of China 14: 58. http://flora.huh.harvard.edu/china/PDF/PDF14/Conium.pdf. (Accessed 15 Oct 2007).

Parmar, C. and M.K. Kaushal. 1982. Wild fruits. Kalyani Publishers, New Delhi.

Partomihardjo, T. 1991. *Nyctanthes arbor-tristis* L. In: Lemmens, R.H.M.J. and N. Wulijarni-Soetjipto (Editors). Plant resources of South-East Asia. No. 3. Dye and tannin-producing plants. Pudoc, Wageningen. Pages 97–99.

Phipps, J.B., R.J. O'Kennon and R.W. Lance. 2003. Hawthorns and medlars. Timber Press, Portland.

Polomski, B. and J.M. Scott. Undated. Fuchsia. Clemson University Extension Cooperative Service. http://hgic.clemson.edu/factsheets/hgic1557.htm. (Accessed 23 Jul 2007).

Polunin, I. 1987. Plants and flowers of Singapore. Times Editions, Singapore.

Porcher, M.H. 1995 onwards. Multilingual Multiscript Plant Name Database. (Updated 18 Jul 2007). http://www.plantnames.unimelb.edu.au/Sorting/List_bot.html#dic1.0. (Accessed 23 Jul 2007).

Porter, B.N. 1993. Sacred trees, date palms, and the royal persona of Ashurnasirpal II. Journal of Near Eastern Studies 52: 129–139.

Radford, E. and M.A. Radford. 1961. Encyclopedia of superstitions. Edited and revised by C. Hole. Hutchinson of London.

Radhakrishnan, C. 2003. Good luck plant. The Hindu. http://www.hindu.com/thehindu/mp/2003/08/05/stories/2003080500220300.htm. (Accessed 23 Jul 2007).

Rahajoe, J.S., R. Kiew and J.L.C.H. van Valkenburg. 1999. *Jasminum sambac* (L.) Aiton. In: de Padua, L.S., N. Bunyapraphatsara and R.H.M.J. Lemmens (Editors). Plant resources of South-East Asia. No. 12(1). Medicinal and poisonous plants 1. PROSEA, Bogor. Pages 319–320.

Randløv C., J. Mehlsen, C.F. Thomsen, C. Hedman, H. von Fircks and K. Winther. 2006. The efficacy of St. John's Wort in patients with minor depressive symptoms or dysthymia – a double-blind placebo-controlled study. Phytomedicine 13: 215–221.

Rao, A.N. and Y.C. Wee. 1989. Singapore trees. Singapore Institute of Biology, Singapore.

Rätsch, C. 1992. The dictionary of sacred and magical plants. Translated by J. Baker. ABC-CLIO, Santa Barbara.

Reuters. 2004. Athens basks in year of the olive. Reuters News. 2 Feb 2004. Reuters Limited. http://www.rediff.com/sports/2004/feb/02athens.htm. (Accessed 21 Nov 2007).

Rieger, M. 2006. Peach – *Prunus persica*. http://www.uga.edu/fruit/peach.html (Accessed 30 November 2007).

Rieger, M. 2006. Plums – *Prunus domestica*, *Prunus salicina*. http://www.uga.edu/fruit/plum.html. (Accessed 3 Dec 2007).

Robertson, K.R. and Y.T. Lee. 1976. The genera of Caesalpinioideae (Leguminosae) in the southeastern United States: 8. *Cercis* Linnaeus. Journal of the Arnold Arboretum 57: 48–53.

Roja, G. and M.R. Heble. 1995. Castanospermine, an HIV inhibitor from tissue cultures of *Castanospermum australe*. Phytotherapy Research 9: 540–542.

Royal Horticultural Society. Undated. Plant of the Month: May: Cercis siliquastrum. http://www.rhs.org.uk/whatson/gardens/wisley/archive/wisleypom03may.asp. (Accessed 5 Nov 2007).

Saxena, R.S., B. Gupta and S. Lata. 2002. Tranquilizing, antihistaminic and purgative activity of *Nyctanthes arbor tristis* leaf extract. Journal of Ethnopharmacology 81: 321-325.

Saxena, R.S., B. Gupta, K.K. Saxena, V.K. Strivastava and D.N. Prasad. 1987. Analgesic, antipyretic and ulcerogenic activity of *Nyctanthes arbor tristis* leaf extract. Journal of Ethno-pharmacology. 19: 193-200

Scheper, J. 1998. Poinsettia. Floridata. (Updated 2004). http://www.floridata.com/ref/E/euph_pul.cfm. (Accessed 4 Aug 2007).
Schmidt, R.J. 1994–2007. Rosaceae (Rose Family). The Botanical Dermatology Database. http://bodd.cf.ac.uk/BotDermFolder/BotDermR/ROSA.html. (Last amended Sep 2007). (Accessed 4 Oct 2007).
Seaton, B. 1995. The language of flowers: a history. University Press of Virginia, Charlottesville and London.
Sievers, A.F. 1930. United States Department of Agriculture Miscellaneous Publication No. 77. The herb hunters guide: American medicinal plants of commercial importance. USDA, Washington D.C. http://www.hort.purdue.edu/newcrop/HerbHunters/hhunters.html. (Accessed 15 Oct 2007).
Silba, J. 1996. Noteworthy conifers in the Xian Shan region of Beijing, China. American Conifer Society Bulletin 13:8–12.
Simoons, F.J. 1998. Plants of life, plants of death. University of Wisconsin Press, Wisconsin.
Skinner, M.F. Undated. 26. *Lilium* Linnaeus, Sp. Pl. 1: 302. 1753; Gen. Pl. ed. 5, 143. 1754. Flora of North America 26: 172. http://www.efloras.org/florataxon.aspx?flora_id=1&taxon_id=118558. (Accessed 27 Oct 2007).
Sotto, R.C. 1992. x*Citrofortunella microcarpa* (Bunge) Wijnands. In: Verjeij, E.W.M. and R.E. Coronel (Editors). Plant resources of South-East Asia. No. 2. Edible fruits and nuts. PROSEA, Bogor. Pages 117–119.
South African National Biodiversity Institute. Undated. Flower emblems of the world. http://www.plantzafrica.com/miscell/floweremblems.htm. (Accessed 25 Jul 2007).
Stearn, W.T. 1996. Stearn's dictionary of plant names for gardeners. Cassell Publishers Limited, London.
Stearn, W.T. and P.H. Davis. 1984. Peonies of Greece. Goulandris Natural History Museum, Kifissia.
Stevens, P. F. 2001 onwards. Angiosperm Phylogeny Website. Version 7, May 2006 [and more or less continuously updated since]. (Updated 13 Jun 2007). http://www.mobot.org/MOBOT/research/APweb/. (Accessed 14 Jul 2007).
Stutley, M. 1985. The illustrated dictionary of Hindu iconography. Routledge and Kegan Paul, London.
Sudiarto and M.A. Rifai. 1992. *Punica granatum* L. In: Verheij, E.W.m. and R.E. Coronel (Editors). Plant resources of South-East Asia. No. 2. Edidble fruits and nuts. PROSEA, Bogor. Pages 270-272.
Sunarto, A.T. 1994. *Ocimum americanum* L. In: Siemonsma, J.S. and K. Piluek (Editors). Plant resources of South-East Asia. No. 8. Vegetables. PROSEA, Bogor, Indonesia. Pages 218–220.
Sze, M.-M. (Editor and translator). 1963. The mustard seed garden manual of painting (Chieh Tzu Yuan Huan Chuan, 1679–1701. A facsimile of the 1887–1888 Shanghai edition translated and edited. With new material included from the 1956 translation. Princeton University Press, Princeton.
Tan, Y. 1991. Flowers: Malaysian meanings. Quill Publishers, Kuala Lumpur.
Tergit, G. 1961. Flowers through the ages. Oswald Wolff, London.
The Columbia Encyclopedia, 6th Edition. 2001–05a. Hyacinth, in botany. Columbia University Press, New York. http://www.bartleby.com/65/hy/hyacinth.html . (Accessed 24 Jun 2007).
The Columbia Encyclopedia, 6th Edition. 2001–05b. Hyacinth, in Greek mythology. Columbia University Press, New York. http://www.bartleby.com/65/hy/Hyacinth.html. (Accessed 24 Jun 2007).
The Herb Society. (2007). Herb fact sheet: Bay. http://www.herbsociety.org.uk/schools/factsheets/bay.htm. (Accessed 8 Oct 2007).
The United States National Arboretum. 2006a. State trees & state flowers. http://www.usna.usda.gov/Gardens/collections/statetreeflower.html. (Accessed 25 Jul 2006).
The United States National Arboretum, 2006b. USDA plant hardiness zone map. http://www.usna.usda.gov/Hardzone/index.html. (Accessed 23 Aug 2007).
The University of Illinois Extension. 2007. Poinsettia facts. http://www.urbanext.uiuc.edu/poinsettia/facts.html. (Accessed 4 Aug 2007).
Thiselton-Dyer, T.F. 2004. The Folk Lore of Plants. Kessinger Publishing.
Tille, A. 1892. German Christmas and the Christmas tree. Folklore 3: 166–182.
Too, L. 1999. Lillian Too's easy-to-use feng shui: 168 ways to success. Collins & Brown, London.
Too, L. 2002. Complete illustrated guide to feng shui for gardeners. Element Books Limited, London.
Upadhyaya, K.D. 1964. Indian botanical folklore. Asian Folklore Studies 23: 15–34.
van der Vossen, H.A.M., G.N. Mashungwa and R.M. Mmolotsi. 2007. *Olea europaea* L. [Internet] Record from Protabase. van der Vossen, H.A.M. and G.S. Mkamilo (Editors). PROTA (Plant Resources of Tropical Africa / Ressources végétales de l'Afrique tropicale), Wageningen, Netherlands. http://database.prota.org/search.htm. (Accessed 20 November 2007).
Vergara, B.S. and S.K. de Datta. 1996. *Oryza sativa* L. In: Grubben, G.J.H. and Soetjipto Paertohardjono. (Editors). Plant resources of South-East Asia. No. 10. Cereals. PROSEA, Bogor. Pp. 106–115.

Vickery, R. (compiler). 1995. A dictionary of plant-lore. Oxford University Press, Oxford.
Vuokko, S. 2001. Finland's national nature symbols. Ministry of Foreign Affairs Finland, Finland. http://virtual.finland.fi/netcomm/news/showarticle.asp?intNWSAID=25582. (Accessed 9 Nov 2007).
Wagner, W.L., D.R. Herbst and S.H. Sohmer. 1990. Manual of the flowering plants of Hawai'i. Volumes 1 and 2. University of Hawaii Press and Bishop Museum Press, Honolulu.
Wee, Y.C. 2003. Tropical trees and shrubs: a selection for urban plantings. Sung Tree Publishing Limited, USA.
Wen, X. 2004 onwards. Interactive Keys by Xiangying Wen. *Camellia japonica*. http://www.efloras.org/florataxon.aspx?flora_id=1001&taxon_id=242310016 and http://www.efloras.org/florataxon.aspx?flora_id=1001&taxon_id=242417714. (Accessed 31 Jul 2007).
Webber, F.R. & Cram, R.A. 2003. Church symbolism. Kessinger Publishing, Whitefish, Montana, USA.
Welman, M. 2005. *Lagenaria siceraria*. http://www.plantzafrica.com/plantklm/lagensic.htm. (Accessed 25 July 2007).
Wichman, T., G. Knox, E. Gilman, D. Sandrock, B. Schutzman, E. Alvarez, R. Schoellhorn and B.
Larson. 2006. Florida-friendly plant list. University of Florida, IFAS Extension, Florida.
Widjaja, E.A. and M.E.C. Reyes. 1994. *Lagenaria siceraria* (Molina) Standley. In: Siemonsma, J.S. and K. Piluek (editors). Plant resources of South-East Asia. No. 8. Vegetables. PROSEA, Bogor, Indonesia. Pages 190–192.
Wong, K.M. 2004. Bamboo: The Amazing Grass. International Plant Genetic Resources Institute and University of Malaya, Kuala Lumpur.
Wong, W. 2007a. Demystifying the "Magic Bean" Part 1. http://www.nparks.gov.sg/blogs/garden_voices/index.php/2007/08/16/demystifying-the-magic-bean/. (Accessed 10 November 2007).
Wong, W. 2007b. Demystifying the "Magic Bean" Part 2. http://www.nparks.gov.sg/blogs/garden_voices/index.php/2007/09/13/demystifying-the-magic-bean-part-2/. (Accessed 10 November 2007).
Wolverton B.C., Douglas W.L., and Bound, K. 1989. A Study of Interior Landscape Plants for Indoor Air Pollution Abatement. An interim report submitted to NASA. http://ntrs.nasa.gov/archive/nasa/ssctrs.ssc.nasa.gov/indr_landscape2/indr_landscape2.pdf. (Accessed 8 Jul 2007).
Wright, A.R. 1903. Some Chinese folklore. Folklore 14: 292–298.
Zhou, T., L. Lu, G. Yang and I.A. Al-Shehbaz. 2001. Brassicaceae (Cruciferae). Flora of China 8: 1–193.

auspicious plants 2-3, 4-5, 8-9, 10-11, 12-13, 14-15, 16-17, 18-19, 20-21, 24-25, 30-35, 36-37, 38-39, 40-41, 42-43, 44-45, 46-47, 48-49, 50-51, 52-53, 54-55, 56-57, 58-59, 60-61, 62-63, 64-65, 66-67, 68-69, 70-71, 72-73, 74-75, 76-77, 78-83, 84-85, 86-87, 88-89, 90-91, 92-93, 94-95, 96-97, 98-99, 102-103, 104-105, 106-107, 108-109, 110-111, 116-117, 118-119, 122-123, 124-125, 126-127, 128-129, 130-131, 132-133, 134-135, 136-137, 142-143, 144-145, 146-149, 150-151, 156-157, 158-159, 160-161, 164-165, 166-167, 170-171, 178-179

British folklore plants 15, 47, 53, 82-83, 91, 97, 99, 109, 153, 157, 163, 173

Buddhist belief plants 41, 75, 111, 119, 151, 165, 171

Chinese folklore plants 9, 39, 57, 65, 77, 89, 105, 119, 145

Chinese New Year plants 9, 17, 21, 25, 37, 39, 41, 43, 57, 63, 67, 77, 85, 87, 89, 95, 105, 127, 129, 131, 143, 145, 161, 167, 179

Christian belief plants 29, 53, 71, 82-83, 91, 93, 95, 103, 109, 125, 133, 141, 151, 153, 161, 163, 177

Christmas plants 30, 35, 67, 91, 95, 99, 175

climbers 20, 88, 90, 100, 104, 138-139, 152, 176

common family names
aroid 178, 180
Australian tree fern 36
banana 116
bean 20, 24, 28, 78, 80, 81, 164, 172
birch 14
bromeliad 8
butcher's broom 60, 62
cactus 64
carnation 58
cashew nut 110
celery 46
citrus 4, 38, 40, 42, 76
cotton 26,128
cucumber 104
custard apple 18
cypress 102, 136, 168
ginger 54
grape 178
grass 12, 92, 126, 158, 170
holly 98
hyacinth 94
iris 86
ivy 90
jade plant 50
laurel 106
lily 108
loosestrife 150
lotus 118
magnolia 112, 114
mahogany 10
mint 122, 156
mulberry 68, 70, 72, 74
mustard 22
nightshade 166
olive 100, 120, 124
palm 44, 132
paperlily 48
peony 130
periwinkle 2, 6
pine 134
rose 52, 142, 144, 146, 152
sandalwood 174
spurge 66
St. John's wort 96
sunflower 56
tea 16
willow 160, 162
willowherb 84
wood sorrel 79

common species names, Catalan
pi blanc 30
pi royal 30

common species names, English
agave 138
American *arbor-vitae* 168
American chestnut 128
American cuckoo flower 22
apple of Sodom 166
Arabian jasmine 100
arbor-vitae 168
aroid palm 178
artificial Christmas trees 30
arum lily 180
ashok tree 164
ashoka 164
ashoka tree 164
Asian broom grass 170
Asian rice 126
asok 164
asoka 164
asoka tree 164
Australian chestnut 24
baby jade 50
badman's oatmeal 46
bael 4
bael fruit 4
bael tree 4
balm-of-Gilead fir 30
balsam fir 30
Baltic redwood 30
bamboo 12
bamboo grass 170
banana 116
Barbary matrimony vine 138
barley 92
bay laurel 106
bay leaf laurel 106
bay tree 106
bead tree 10
beaver poison 46
belfruit tree 4
Belgian evergreen 62
Bengal quince 4
Benjamin fig 68
Benjamin tree 68
Benjamin's fig 68
birch 14
bird's eye chilli 88
black bean 24
black medick 78
black trefoil 78
blackboard tree 6
blister fir 30
blur spruce 30
bo tree 74
bodhi tree 74
border carnation 58
bottle gourd 104
bouquet grass 170
broad bean 172
Buddha's hand 40
Buddha's hand citron 40
bullet kumquat 76
bunk 46
Burmese neem tree 10
cacti 138
cactus family members 138

calabash 104
calabash gourd 104
calamandarin 38
calamondin 38
calamondin orange 38
calamonding 38
calla lily 180
camellia 16
Canada balsam fir 30
carnation 58
cashes 46
Caucasian fir 30
cauliflower-ear's 50
century plant 138
champa 114
champac 114
champaca 114
champak 114
champaka 114
cherry blossom 146
chili pepper 88
chilli 88
China orange 38
Chinaberry 10
Chinese *arbor-vitae* 136
Chinese banyan 72
Chinese boxthorn 138
Chinese matrimony vine 138
Chinese orange 38
Chinese plum 142
Chinese water bamboo 62
Chinese wolfberry 138
Christ's thorn 138
Christ's thorn jujube 138
Christmas flower 66
Christmas plant 66
Christmas star 66
chrysanthemum 56
clove pink 58
coco palm 44
coconut 44
coconut palm 44
Colorado blue spruce 30
common arum lily 180
common barley 92
common calla 180
common calla lily 180
common camellia 16
common fig 70
common grapevine 176
common hyacinth 94
common ivy 90
common mandarin 42
common mistletoe 174
common peach 144
common pine 30
common rice 126
common wood sorrel 79
compact dracaena 60
compass plant 156
copal 26
coral jasmine 120
corn plant 60
corn-flag 86
cornstalk plant 60
cotton thistle 138
cotton tree 26
cotton wood 26
cow's udder 166
cow's udder plant 166
crown of thorns 138
cuckoo bread 79
cuckoo bittercress 22
cuckoo flower 22
culate mandarin 42
cultivated barley 92
cultivated fig 70
cultivated rice 126
cumquat 76
cumquot 76
curcuma 54
curly bamboo 62
curtain fig 72
date 132
date palm 132
desert rose 2
devil tree 6
Devil's oatmeal 46
dita bark 6
divine flower 58
dollar plant 50
Douglas fir 30
Douglas red spruce 30
Douglas yellow spruce 30
Douglastree 30
dry date 132
Duke of Argyll's tea tree 138
Dutch clover 81
Dutch hyacinth 94
Dutchman's pipe 64
Dutchman's pipe cactus 64
dwarf ambarella 88
East Asian cherry 146
East Indian lotus 118
Eastern *arbor-vitae* 168
eastern fir 30
Eastern white cedar 168
edible date 132
edible fig 70
Egyptian lily 180
Egyptian lotus 118
elephant apple 4
English bean 172
English ivy 90
eternity plant 178
euphorbia 138
European bean 172
European grape 176
European mistletoe 174
European silver fir 30
European wood-sorrel 79
faba bean 172
false palm 48
fat boy 178
fava bean 172
field bean 172
fig 70
fingered citron 40
flame leaf flower 66
flesh finger citron 40
florist's calla 180
florist's chrysanthemum 56
fragrant champaca 114
fragrant dracaena 60
fragrant Himalayan champaca 114
Fraser fir 30
friendship bamboo 62
friendship bush 156
friendship tree 50
fuchsia 84
garden calla 180
garden chrysanthemum 56
garden hyacinth 94
garden lily 108
garden mum 56
garden rose 152
garden tomato 88
garland thorn 138
giant white arum lily 180
gladiolus 86
goat pepper 88
goat's willow 160
Goddess of Mercy plant 62
goji berry 138
golden apple 4, 88
golden champaca 114
golden fig 68
golden lime 38
golden orange 76
golden rooster 36
good luck plant 48
grape 176
grapevine 176
great hog plum 88
great sallow 160
Grecian laurel 106
green corn plant 60
green snob 79
Guiana chestnut 128
haw 52
hawthorn 52
heck-how 46
hemlock 46
heraldic thistle 138
hog plum 88
holly 98
holy basil 122
honeysuckle clover 81
hooded barley 92

hop clover 78
horse bean 172
hot chili 88
hyacinth 94
hypericum 96
impala lily 2
Indian cedar 10
Indian devil tree 6
Indian fig tree 74
Indian laurel fig 72
Indian lilac 10
Indian lotus 118
Indian mango 110
Indian pulai 6
Indian quince 4
Indian saffron 54
Irish shamrock 79
ivy 90
Jack in the pulpit 180
jade plant 50
jade tree 50
Japanese apricot 142
Japanese camellia 16
Japanese laurel 50
Japanese plum 142
Japanese rubber plant 50
jasmine 100
Java fig 68
Java tree 68
Jerusalem cherry 88
Jerusalem thorn 138
Jew plum 88
Judas tree 28
juniper 102
kalamansi lime 38
kapok 26
kapok tree 26
ki 48
kinkan 76
knotty pachira 128
kumquat 76
kuntz 170
ladino clover 81
lady of the night 64
lady's eardrops 84
lady's smock 22
laurel 106
laurel fig 72
lesser trefoil 80
lesser yellow trefoil 80
lily 108
lily of the Nile 180
limequat 38
little hop clover 80
lobster plant 66
long rooted curcuma 54
lotus 118
love apple 88, 166
love tree 28
low hop clover 80
lowland rice 126
lucky bamboo 62
lucky bean 24
lucky bean plant 24
macaw bush 166
Madeira cherry 88
Madeira winter cherry 88
makopa 88
Malabar chestnut 128
Malay banyan 72
Malayan banyan 72
malting barley 92
mandarin 42
mandarin orange 42
mango 110
margosa 10
margosa tree 10
marumi kumquat 76
matrimony vine 138
may flower 22
meadow cress 22
mei 142
Mexican flame leaf 66
michelia 114
milkwood pine 6
milky pine 6
money plant 128
money tree 50, 128, 178
monks' basil 122
Moreton Bay chestnut 24
mum 56
mume 142
musk lime 38
nagami kumquat 76
nariyal 44
neem 10
neem tree 10
New Zealand Christmas tree 30
night blooming cereus 64
night jasmine 120
night-blooming jasmine 120
night-flowering jasmine 120
nimtree 10
nipple fruit 166
noble cane 158
noble fir 30
noble sugar cane 158
nonesuch 78
Nordmann fir 30
Northern white cedar 168
Norway spruce 30
olive 124
orange chempaka 114
Oregon pine 30
Oriental *arbor-vitae* 136
oriental cherry 146
Oriental lotus 118
oriental thuja 136
Otaheite apple 88
oval kumquat 76
paddy rice 126
paeony 130
painted leaf 66
palm lily 48
Panama orange 38
Parry spruce 30
peach 144
peach blossom 144
pearl lemon 76
peepul tree 74
peony 130
perfume tree 18
Philippine lime 38
pig lily 180
pig's ears 166
pine 134
pineapple 8
pipal tree 74
pippala 74
pipul 74
plant of friendship 156
plum blossom 142
po tree 74
pohutukawa 30
poinsettia 66
poison hemlock 46
poison parsley 46
poison root 46
poison snakeweed 46
Polynesian plum 88
Polynesian ti plant 48
pomegranate 150
Potomac cherry 146
provision tree 128
pulai 6
pungent pepper 88
pussy willow 160
queen of the night 64
red basil 122
red date 132
red fir 30
red pine 30
red suckling clover 80
ribbon dracaena 62
ribbon plant 62
rice 126
Roman hyacinth 94
rose 152
rosemary 156
rough basil 122
round kumquat 76
saba nut 128
sacred basil 122
sacred bean of India 118
sacred fig 74
sacred fig tree 74
sacred lotus 118
sacred Siamese basil 122
sacred Thai basil 122
sacred water lily 118

sad tree 120
Sander's dracaena 62
santara orange 42
sapu 114
scarlet lime 38
Scotch pine 30
Scotch thistle 138
Scots pine 30
Scottish thistle 138
Scythian lamb 36
semisoft date 132
shamrock 79, 80, 81
silk cotton tree 26
silver spruce 30
sleeping beauty 79
small hop clover 80
soft date 132
sorrel 79
sorrowless tree 164
southern balsam fir 30
spicy-peeled kumpuat 76
spinks 22
spotted conium 46
spotted cowbane 46
spotted hemlock 46
spotted parsley 46
spur pepper 88
St. Bennet's herb 46
St. John's wort 96
stone apple 4
striped dracaena 60
suckling clover 80
sugarcane 158
suntara orange 42
sweet bay 106
sword bean 20
sword flag 86
sword jackbean 20
sword lily 86
Syrian Christ's thorn 138
Tahitian quince 88
tangerine 42
Tartarian lamb 36
Thai basil 122
thorn 52
thorn apple 52
thorny burnet 138
ti 48
ti plant 48
tiger grass 170
titty fruit 166
Tokyo cherry 146
tomato 88
tree of kings 48
tree of sadness 120
tree of sorrow 120
trefoil 78
tropic laurel 68
true kapok tree 26
true mandarin 42
trumpet lily 180
turmeric 54
upland rice 126
vine 176
water bamboo 62
water chestnut 128
weeping Chinese banyan 68
weeping fig 68
weeping laurel 68
weeping willow 162
Western Australian Christmas tree 30
white arum lily 180
white cedar 168
white champaca 112
white cheesewood 6
white chempaka 112
white clover 81
white Dutch clover 81
white sandalwood 112
white silk cotton tree 26
white spruce 30
white wood sorrel 79
white-flowered gourd 104
whitethorn 52
Whitsun flower 79
wi tree 88
wild fig 70
willow 162
Windsor bean 172
winter cherry 88
winter rose 66
wode whistle 46
wolfberry 138
wood apple 4
wood sorrel 79
woolly fern 36
woolly thistle 138
yellow ginger 54
yellow jade orchid tree 114
yellow plum 88
yellow trefoil 78
ylang-ylang 18
Yoshino cherry 146
zambac 100
zombi apple 166
zz plant 178

common species names, French

abre à kapok 26
abre bo de 74
abre coton 26
abre kapok 26
abricot du Japon 142
abricotier du Japon 142
abricotier Japonais 142
agave 138
amliso, herbe à balai 170
ananas commun 8
arbre ashoka 164
arbre de l'Intendance 72
arrow-root de l'Inde 54
arum 180
azadirachta de l'inde 10
bambou 12
bambou de'eau 62
bambou porte-bonheur 62
banane 116
basilic sacré 122
basilic sacré à feuilles vertes 122
basilic Thaïlandais 122
bel Indien 4
bois coton 26
cacaoyer-rivière 128
cactus 138
calamondin 38
calebassier 104
calebassier grimpant 104
camélia du Japon 16
camellia 16
canang odorant 18
canne à sucre 158
canne Chinoise 62
canne de Chine 62
capoc 26
capoquier 26
casamangue 88
cédrat digité 40
cédrat main de Bouddha 40
cerisier à feuilles en dents de scie 146
cerisier du collines 146
cerisier du Japon 146
cerisier Yoshino 146
chardon aux ânes 138
châtaigner sauvage 128
châtaignier de la Guyane 128
chrysanthème 56
ciguë 46
cocotier 44
cotonnier de l'Inde 26
cougourde 104
courge massue 104
courge siphon 104
cresson des prés 22
cucurma 54
cucurma long 54
dattier 132
dattier commun 132
de Douglas 30
dolic en sabre 20
dolique sabre 20
dracaena de Chine 62
epiciéa commun 30
épinette bleue 30
épinette de Norvège 30
epiphyllum à large feuilles 64
étoile de Noël, poinsettia 66

euphorbes 138
faux cotonnier 26
fève 172
fève a cheval 172
fève d'Egypte 118
fève de cheval 172
fève des marais 172
féverole 172
figuier comun 70
figuier de l'Inde 74
figuier des banians 74
figuier des pagodas 74
figuier Indien 74
figuier sacré 74
fromager 26
fuchsia 84
genièvre 102
glaïeul 86
gorgane 172
gourde bouteille 104
gourgane 172
grappe 176
grenade 150
grenadier 150
grosse fève 172
grosse fève commune 172
gui 174
haricot sabre 20
houx 98
ilang ilang 18
incensier 156
jacinthe 94
jasmin, jasmin d'Arabie 100
kapokier 26
kapokier de Java 26
kumquat â chair acide 76
kumquat â fruit oblongs 76
kumquat â fruits ronds 76
kumquat ovale 76
kumquat rond 76
l'arbre de Judée 28
laurier 106
lierre 90
lilas des Indes 10
lis 108
lotus des Indes 118
lotus du Nil 118
lotus Indien 118
lotus magnolia 118
lotus sacré 118
lotus sacré de l'Inde 118
lupuline 78
luzerne lupuline 78
lyciet de Chine 138
lys 108
main de Bouddha 40
mandarine 42
mandarinier 42
mangue 110
margosier 10
margosier de Birmanie 10
margosier du Bangladesh 10
millepertuis 96
minette 78
minette dorée 78
neem des Indes 10
nim des Indes 10
noisetier de la Guyane 128
noix de coco 44
oeillet 58
olive 124
oranger du Malabar 4
orge 92
orge d'hiver 92
orge de printemps 92
orge vulgaire 92
pachirer aquatique 128
palmier dattier 132
pêche 144
pêcher 144
pignon 134
piment 88
piment des oiseaux 88
piment enragé 88
pin 134
pin commun 30
pin d'Auvergne 30
pin d'Ecosse 30
pin de Genève 30
pin de Haguenau 30
pin de Norvège 30
pin de Riga 30
pin de Russie 30
pin rouge du nord 30
pin sylvestre 30
pinasse 30
pivoine 130
plant zz 178
pois de l'Inde 20
pois sabre 20
pois sabre de la Jamaïque 20
pois sabre rouge 20
pomme de cythère 88
pomme zombi 166
pommier de cythère 88
prune d'ume 142
prune de cythère 88
prune du Japon 142
prunier d'Amérique 88
prunier de cythère 88
prunier Japonais 142
raisin 176
ris commun 126
riz 126
riz cargo 126
riz cultivé 126
riz de plaine 126
riz non décortiqué 126
riz paddy 126
riz vétu 126
romarin 156
romarin commun 156
rose 152
rose du Japon 16
sacrcodactyle 40
sacré de bodh-gaya 74
safran des Indes 54
saule 162
saule blanc 160
saule marsault 160
tetons de jeune fille 166
thuya d'Occident 168
thuya du Canada 168
thuya le Chine 136
tomate 88
trèfle douteux 80
trèfle etalé 80
trèfle filiforme 80
umé 142
vigne 176
vigne cultivée 176
ylang ylang 18

common species names, German

Abendländische Lebensbaum 168
Abendländische Thuja 168
Ackerbohne 172
Adventsstern 66
Agave 138
Ashokbaum 164
Bambus 12
Banane 116
Baum 74
Baumwollbaum 26
Belbaum 4
Bengalische Quitte 4
Benjamin-Gummibaum 68
Birke 14
Birkenfeige 68
Bobaum 74
Bodhi-Baum 74
Borksdorn 138
Buddhafinger 40
Chili Pfeffer 88
Chinesische Feige 72
Chinesische Kumquat 76
Chinesischer Borksdorn 138
Chinessischer Pflaumenbaum 142
Christstern 66
Chrysantheme 56
Dattelpalme 132
Dicke Bohne 172
Dreibluetige Pflaume 142
Echte Feige 70
Echter Weinstock 176
Efeu 90
Eselsdistel 138

Euter Nachtschatten 166
Faden Klee 80
Flaschenkürbis 84
Flaschen-Kürbis 84
Föhre 30
Forche 30
Forle 30
Fuchsie 84
Fuma 26
Gefleckter Schierling 46
Gelber Wiesen-Klee 80
Gelbklee 78 Gelbwurz 54
Gelbwurzel 54
Gemein Fichte 30
Gemeine Föhre 30
Gemeine Kiefer 30
Gemeiner Reis 126
Gerfingerte Zitrone 40
Gerste 92
Gesägtblättrige Kirsche 146
Gewöhnliche Judasbaum 28
Gewöhnliche Wald-Kiefer 30
Gewöhnlicher Flaschenkürbis 84
Gilber Ingwer 54
Gilbwurzel 54
Gladiole 86
Goldpflaume 88
Granatapfel 150
Grannenkirsche 146
Hartheu 96
Heiliger Feigenbaum 74
Hopfenklee 78
Hyazinthe 94
Ilang Ilang 18
Indische Quitte 4
Indischer Basilikum 122
Indischer Lorbeer 72
Indischer Lotus 118
Indischer Mango 110
Indischer Safran 54
Japanische Aprikose 142
Japanische Blütenkirsche 146
Japanische Kamlie 16
Japanische Kirsche 146
Japanischer Aprikosenbaum 142
Japanischer Pflaumenbaum 142
Jasmin 100
Jerusalemkirsche 88
Johanniskräuter 96
Kaktus 138
Kalebassenkürbis 84
Kamelie 16
Kapokbaum 26
Keulenlilie 48
Kiefer 134
Kokos 44
Kokosnuß 44
Kokospalme 44
Korallenstrauch 88
Kulturgerste 92
Kumquat 76
Kurkuma 54
Lilie 108
Lorbeer 106
Lorbeerfeige 72
Mandarine 42
Mandarinenbaum 42
Mango 110
Mistel 174
Mombinpflaume 88
Morgenländische Lebensbaum 136
Mumebaum 142
Nelke 58
Nelkenkirsche 146
Niembaum 10
Nimbaum 10
Nordische Kieefer 30
Olive 124
Oval Kumquat 76
Ovale Kumquat 76
Ovaler Kumquat 76
Paddy-Reis 126
Panaschierter Drachenbaum 62
Päonie 130
Pappelfeige 74
Pepul-Baum 74
Pepulbaum de Inder 74
Pferdebhne 172
Pfirsich 144
Poinsettie 66
Puffbohne 172
Quitte 4
Rebe 176
Rebstock 176
Reis 126
Rohreis 126
Rose 152
Rosmarin 156
Roter Pfeffer 88
Rundkumquat 76
Saatsgerste 92
Salweide 160
Saubohne 172
Saubohnen 172
Schierling 46
Schneeaprikose 142
Schwertbohne 20
Stechpalme 98
Süsse Mombinpflaume 88
Tahitia-Apfel 88
Tahitiapfel 88
Tokyokirsche 146
Tomate 88
Trompetenkürbis 104
Vogelpfeffer 88
Wacholder 102
Wald-Föhre 30
Waldkiefer 30
Weide 162
Weidenkätzchen 160
Weihnachtsstern 66
Wein 176
Wein-Rebe 176
Weinrebe 176
Weinstock 176
Weisenschaumkraut 22
Wilder Cacaobaum 128
Wilder Kakaobaum 128
Wolfsmilch 138
Wollbaum 26
Yoshinokirsche 146
Ziegenpfeffer 88
Zitzen Nachtschatten 166
Zuckerrhor 158
Zweifelhafter Klee 80
Zwergapfelsine 38
Zwerg-Klee 80

common species names, Hindi

akasha beli 174
am 110
amara bela 174
amra 88, 110
anaarlanar 150
anannaasa 8
anhuri 172
anjir 70
aru 144
ashok 164
asok 164
baans 12
baansa 12
bakla 172
balnimb 10
baranda 122
bela 100
beli 4
beligri 4
bhurj 14
bohj 14
chameli 100
chavala 126
chilkan 72
daudi 56
dhana 126
dudhi 104
ganna 158
gannaa 158
gul 152
guladaudi 56
hab-el-gar 106
haldi 54
haldii 54
harsinghar 120
hauber 102

iikh 158
jaituna 124
jalapai 124
jangalii tulasii 122
jau 92
kaalaa tulasii 122
kaiktasa 138
kala matar 172
kamal 118
kamala 108
kamarup 72
kanwal 118
katan 26
kelaa 116
khaji 132
khajur 132
khopar 44
Krishna tulasii 122
kumuda 108
kumuda 118
lalmirch 88
lamirchi 88
lankamirchi 88
lokhi 104
magadhi 100
mirch 88
nalini 108
naragisa 108
nariyal 44
nariyal kaa per 44
neem 10
nim 10
nind 10
padma 118
pattra 14
pipal 74
pipali 74
pipli 74
rusmari 156
safed savara 26
safed semul 26
safed simal 26
safed simul 26
sakhara 158
salma 132
santara 42
sela 126
sendhi 132
sephalika 120
shaukaran 46
sirphal 4
sita ashok 164
sumbul 94
tamatar 88
tulasii 122
tulsi 122
ukh 158
varanda 122
vrikshalata 90
zaitun 124

common species names, Japanese
ke momo 144
miume 142
momo 144
piichi 144
sakura 146
sato zakura 146
somei yoshino 146
sumomo 142
ume 142
ume no mi 142
yama zakura 146

common species names, Malay/Indonesian
ampelam 110
anggur 176
anjir 70
aur 12
bakawali 64
bakung 108
barli 92
beras Belanda 92
beringin 68
bila 4
bilak 4
birch 14
bodi 74
buah anggur 176
buah delima 150
buloh 12
bulu empusi 36
bulu pusi 36
buluh 12
buluh rumput 170
buluh tebrau 170
bunga kekwa 56
bunga padam 118
bunga sakura 146
bunga telepok 118
cabai 88
cabai burong 88
cabai Melaka 88
cabai merah 88
cabai rawit 88
cempa 114
cempaka 112
cempaka gading 112
cempaka merah 114
cempaka putih 112
cenanga 18
ceri 146
cili 88
cili padi 88
dedalu 162
dedalu 174
dedalu kambing 160
delima 150
euphorbia 138
gapis 164
hemlok 46
holi 98
ijas jepang 142
intaran 10
ivy 90
jampaka 114
jawi jawi 68
jejawi 68
jejawi 72
jeluang 48
kabu kabu 26
kacang babi 172
kacang parang 20
kacang polong 20
kakantrie kapok 26
kaktus 138
kapok abu 26
kapok kapok 26
kastuba 66
kauki 138
kedongdong 88
kekabu 26
kelambir 44
kelapa 44
kelat sega 68
kemangi 122
kenanga 18
kenanga hutan 18
kerambil 44
koki 138
kunyit 54
kurma 132
lada api 88
lada merah 88
laurel 106
limau jari 40
limau kerat lintang 40
limau kupas 42
limau langkat 42
limau wangkang 42
mambu 10
mangga 110
melati 100
melur 100
mempelam 110
mengkapas 26
mind 10
nanas 8
nasi 126
nyiur 44
padema 118
padi 126
padi Belanda 92
pain 134
penawar jambi 36
persik 144
pisang 116
poinsettia 66
pokok kapas 26

pokok kenanga 18
pulai 6
randu 26
ros 152
sadu 10
sawang 48
selasih merah 122
selasih Siam 122
semenderasa 18
seri gading 120
seroja 118
talan 164
tamar 132
tanglin 164
tebu 158
tengalan 164
teratai 118
terung masam 88
thistle 138
tomato 88
waringin 68
willow 162
zait 124
zaitun 124

common species names, Mandarin

bái guǒ hú jì shěng 174
bǎi hé 108
bái róng 68
bái yù lán 112
běi měi xiāng bǎi 168
bīn zǎo 138
bīng láng qīng 88
bō luó 8
bù bù gāo 62
cán dòu 172
cǎo diàn suì mǐ jì 22
cè bǎi 136
chá huā 16
cháng chūn téng 90
chúi liǔ 162
chuí yè róng 68
cì bái 102
dà chì jì 138
dà jì 138
dà mài 92
dà táo rén 144
dà zòng yè lú 170
dāo dòu 20
dāo guà jīn zhōng 84
dào zǐ 126
dōng jīng yīng huā 146
dōng qīng 98
dú shén 46
fān qié 88
fèng lí 8
fēng xìn zǐ 94
fó shǒu 40
fó shǒu gān 40
fù guì huā 2
gān 42
gǎn lǎn 124
gān zhè 158
gāo jiā suǒ lěng shān 30
gǔo qǐ 138
hǎi zǎo 132
hé 118
hú lú 104
hú lú guā 104
huà mù 14
huā qí sōng 30
huáng huā liǔ 160
huáng jiāng 54
huáng lán 114
huáng lí 8
huáng shū cǎo 80
huáng yù lán 114
jí bèi 26
jí bèi mù mián 26
jiān lán 86
jiàn zhú máo 170
jié 42
jīn gān 76
jīn jú 76
jīn máo gǒu 36
jīn qián shù 178
jīn sī táo 96
jú 42, 56
jú huā 56
kāng nǎi xīn 58
kǔ liàn 10
kuān pí gān 42
kuān pí jié 42
là jiāo 88
lián 118
liàn shù 10
liàn zào zǐ 10
liǔ 162
lóng shé lán 138
luǎn yè hú jì shěng 174
mǎ tí lián 180
máng guǒ 110
máo táo 144
méi 142
méi gùi 152
mǐ 126
mí dié xiāng 156
mì gān 42
mǐ jiāo 88
mò lì huā 100
mù jié 4
nán ōu zǐ jīng 28
níng xià gǒu qǐ 138
ōu zhōu chì sōng 30
ōu zhōu pú táo 176
ōu zhōu yún shān 30
pèng gān 42
pú táo 176
pú tí shù 74
qiáng 152
qiáng wēi 152
rǎn jǐng jí yě yīng 146
rén miàn zǐ 88
rì běn lǐ 142
róng shù 72
rǔ qié 166
shān chá huā 16
shān hú yīng 88
shān jú 76
shān yīng huā 146
shān zhā 52
sháo yào 130
shè xiāng shí zhú 58
shèng luó lè 122
shí líu 150
shǔi daò 126
sì jì jú 38
sī wéi shù 74
sōng 134
tán huā 64
táng chāng 86
táng jiāo shù 6
táo 144
táo rén 144
táo zǐ 144
tiān lán mù xù 78
tiě shù 60
tú gān 42
wáng lí 8
wú huā guǒ 70
wú méi 142
wú yōu huā 164
wú zhǐ gān 40
xiān rén zhǎng kē 138
xiàn sān yè cǎo 80
xiàn yé sān yè cǎo 80
xiāng jiāo 116
xiāng lóng xuè shù 60
xiāng yè 106
xiǎo là jiāo 88
xìng yùn dòu 24
yà zhōu zāi péi dào 126
yè huā 120
yě jiāo zi 88
yě là zǐ 88
yē shù 44
yē zǎo 132
yē zǐ 44
yī lán 18
yì pǐn hóng 66
yì wèi lóng xuè shù 60
yìn dù liàn shù 10
yín yè lóng xuè shù 62
yù jīn 54
yù shù 50
yuè guì 106
yuè jú 38
zhāo cái shù 128
zhú 12

zhū jiāo 48
zhuā wā mù mián 26
zhuǎn yùn zhú 62
zhuàng lì lěng shān 30

common species names, Russian
jel europeiskaya 30

common species names, Sanskrit
saptaparna 6

common species names, Spanish
abedul 14
abeto 30
acebo 98
aceituna 124
acocote 104
agave 138
aji 88
albaricoquero japonés 142
alfalfa lupulina 78
ambarella 88
araná 8
árbol Benjamin 68
arbol capoc 26
árbol de amor 28
árbol de Judea 28
árbol de la Judas 28
arbol de la seda 26
arbol de seda 26
arbol sagrado de la India 74
arbor-vitae chino 136
arroz 126
arroz con cáscara 126
arroz con cáscara asiático 126
arroz irrigado 126
azafrán de la India 54
bambú 12
banana 116
bela 4 milva 4
Benjamina 68
biota 136
cacao cimarrón 128
cacao de monte 128
cacao de playa 128
cactus 128
cadmia 18
cajombre 104
calabaza 104
calabaza vinatera 104
camelia 16
caña común 158
caña de azúcar 158
caña de castilla 158
caña de indio 48
caña dulce 158
caña melar 158
caña sacarina 158
canaduz 158
cañamiel 158
cananga 18
capoquero 26
carabanz 20
cardo borriquero 138
carretilla 78
carretón 78
castaño 128
castaño de agua 128
castaño de la Guayana 128
castaño silvestre 128
cebada 92
cebada común 92
cebada cultivada 92
ceiba juca 26
ceibo de agua 128
ceibo de arroyo 128
cereza de Jerusalén 88
cereza de Madeira 88
chila blanca 128
chile 88
chile 88
cicuta 46
cirolero japonés 142
ciruela japonésa 142
ciruelo japonés 142
clavel 58
coco 44
cocotero 44
crisantemo 56
croto 48
dátil 132
dracena de Africa tropical 60
dracena de Camerún 62
enebro 102
estoque 86
euphorbia 138
faba 172
fenarola-menuda 78
ficus Benyamina 68
flor de Pascua 66
fucsia 84
gladiola 86
gladiólo 86
granada 150
granado 150
guindilla 88
guiro amargo 104
haba 172
haba caballar 172
haba comun 172
haba de burro 20
haba de huerta 172
haba mayor 172
habichuela 172
haboncillo 172
herba de la desfeta 78
hiedra 90
hierba de San Juan 96
higo 70
higuera común 70
higuera de las pagodas 74
higuera religiosa de la India 74
higuera sagrada de los Budistas 74
hipérico 96
ilang ilang 18
jacinto 94
jazmin 100
jazmin de Arabia 100
jelinjoche 128
jobo de la India 88
kumquat redondo 76
laural 106
laurel de Indias 72
lirio 108
loto sagrado 118
lotus 118
lupulina 78
mandarina 42
mandarino 42
mango 110
mangrano 150
matapolo 68
melgó menut 78
meligón 78
melocotón 144
mielga 78
mielga azafranada 78
mielga negra 78
milva 4
mosmote 26
muérdago 174
naranjita China 76
naranjita de San José 38
naranjita Japonesa 76
nuez de coco 44
oliva 124
palma de coco 44
palmera datilera 132
palmera de coco 44
palmillo 60
palo de boya tetón 128
pamera de dátiles 132
peem 26
peonia 130
pérsico duraznero 144
pimienta picante 88
piña 8
piña de américa 8
piña tropical 8
pino 134
pino albar 30
pino blancal 30
pino común 30
pino de valsain 30
pino norte 30
pino Oregon 30

pino real 30
pino rojal 30
pino royano 30
pino royo 30
pino serrano 30
pino silvestre 30
plátano 116
poinsettia 66
pombo 128
poroto sable 20
pumpunjuche 128
quirihillo 128
reina de la noche 64
romero 156
romero comun 156
rosa 152
rosa del Nilo 118
rosmario 156
sapotolón 128
sauce 162
sauce blanco 160
sauce llorón 162
sunzapote 128
tetilla 166
tomate 88
trébol amarillo 80
trébol filiforme 80
trèvol 78
tsine 128
vino 176
yerba de San Juan 96
yuca 26
zapote bobo 128
zapote de agua 128
zapotolongo 128
zapotón 128

evil spirit plants 6-7, 18-19, 26-27, 46-47, 64-65, 68-69, 72-73, 112-113, 114-115, 116-117

herbs (non-woody plants) 8-9, 22-23, 36-37, 46-47, 50-51, 54-55, 56-57, 58-59, 62-63, 78-83, 86-87, 88-89, 92-93, 94-95, 96-97, 108-109, 116-117, 118-119, 122-123, 126-127, 130-131, 138-141, 158-159, 170-171, 178-179, 180-181

Greek mythology plants 47, 95, 107, 125, 131, 135, 151, 157, 175

Hindu belief plants 5, 7, 11, 45, 55, 75, 101, 111, 113, 115, 117, 119, 121, 123, 159, 165

inauspicious plants 4-5, 6-7, 8-9, 18-19, 22-23, 26-27, 28-29, 44-45, 56-57, 58-59, 64-65, 66-67, 68-69, 70-71, 72-73, 86-87, 90-91, 102-103, 106-107, 108-109, 134-135, 138-141, 150-151, 152-155, 156-157, 158-159, 162-163, 172-173, 180-181

Indian folklore plants 10, 45, 49, 101, 111

Japanese folklore plants 145, 147

Judaism belief plants 71, 93, 125, 133, 151, 177

Malay folkore plants 19, 45, 65, 69, 101, 113, 115, 117, 127

medicinal properties and uses 5, 7, 11, 37, 39, 41, 55, 91, 107, 109, 121

Muslim belief plants 71, 125

natural distribution
Afghanistan 92, 174, 176
Africa 2, 12, 20, 60, 62, 86, 102, 104, 134, 174, 178, 180
America, tropical 26, 158
Arabia, 2, 132
Arctic 102, 134
Asia Minor 106
Asia 12, 20, 38, 46, 56, 78, 90, 118, 130, 172, 174
Australia 6, 18, 20, 24, 30, 48, 68, 72, 116, 118, 180
Bangladesh 4
Black Sea, south of 176
Bonin Islands 72
Brazil 128
Bulgaria 28
California 180
Cambodia 4
Cameroon 62
Canada 168
Carribean 158, 166
Caucasus 80
Central America 12, 84, 102, 128, 134, 166
Central Asia 81
China 22, 36, 38, 40, 56, 68, 72, 76, 114, 126, 136, 142, 144, 146, 158, 170
Christmas Island 72
Cocos Island 72
cosmopolitan 96, 98
Eurasia 86
Europe 22, 30, 46, 78, 80, 88, 94, 130, 160
Fiji 10, 18
Gautemala 64
Great Britain 30, 180
Hawaii 48, 158, 180
Himalayas 48, 102, 124, 150
India 4, 6, 36, 40, 54, 68, 72, 74, 92, 100, 108, 114, 120, 122, 126, 146, 150, 158, 164, 170, 174
Indo-China 40, 110, 114, 116
Indonesia 36, 114, 122, 134
Japan 16, 22, 40, 142, 146, 174
Kazakhstan 22
Kenya 178
Korea 16, 22, 146
Laos 4
Lebanon 28
Macronesia 180
Madagascar 86
Malay Archipelago 18, 158
Malaysia 36, 114, 116, 122, 164, 170
Mascarenes 180
Mauritius 10
Mediterranean 28, 58, 70, 88, 94, 106, 150, 156, 158, 172
Melanesia 48
Mexico 64, 66, 128, 166
Middle East 92
Mongolia 22
Morocco 81
Mozambique 50
Myanmar 4, 10, 18, 110, 164
Nepal 4, 120
New Caledonia 24
New Guinea 18, 48, 158
New Zealand 30, 84, 180
North America 22, 46, 130, 166
North Atlantic island archipelagos 124
Northeast Asia 160
Northern Hemisphere 14, 108, 162
Pacific, the 20, 44
Pakistan 4, 120, 124, 150, 164, 174
Palau 72
Puerto Rico 180
Philippines, the 108, 122
Poland 176
Polynesia 48
Russia 22
Ryukyu Islands 72

Sahara 132
Sikkim 146
Sind 124
Solomon Islands 6, 68, 72
South Africa 50, 124, 178, 180
South America 8, 12, 84, 128, 166
Southeast Asia 6, 38, 42, 48, 68, 72, 74, 118
Southern Hemisphere 162
Southwest Asia 70
Sri Lanka 6, 72, 122, 164
subtropics, the 78, 100, 112
Tahiti 84
Taiwan 16, 102
temperate lands 12, 46, 52, 80, 100, 130, 134, 152
Thailand 120, 126
tropics, the 12, 18, 78, 100, 104, 112, 122
Truk 72
Tunisia 81
Turkey 28, 150
USA 30, 168, 180
Vanuatu 24
Vietnam 4
Volga 118
West Africa 26
West Bengal 146
Western New Britain 24
Yangtze River valley 126

protective aura/power against evil plants 4-5, 8-9, 10-11, 12-13, 14-15, 22-23, 40-41, 44-45, 48-49, 52-53, 60-61, 62-63, 78-83, 86-87, 96-97, 104-105, 152-153

scientific names
Abies alba 30
Abies balsamea 30
Abies fraseri 30
Abies nordmanniana 30
Abies procera 30
Adenium obesum 2-3
Aegle marmelos 4-5
Agave species 138
Alstonia scholaris 6-7
Ananas comosus 8-9
Azadirachta indica (= *Melia indica*, *Melia azadirachta*) 10-11
Bambuseae 12-13
Betula species 14-15
Cactaceae 138
Camellia japonica 16-17
Cananga odorata (= *Canangium odoratum*) 18-19
Canavalia gladiata (= *Canavalia ensiformis* var. *gladiata*) 20-21
Capsicum annuum cultivated varieties 88
Capsicum frutescens 88
Cardamine pratensis 22-23
Castanospermum australe 24-25
Ceiba pentandra 26-27
Cercis siliquastrum 28-29
Christmas tree 30-35
Cibotium barometz 36-37
x*Citrofortunella microcarpa* (= *Citrus madurensis*, *Citrus microcarpa*, *Citrus mitis*, x*Citrofortunella mitis*) 38-39
Citrus medica var. *sarcodactylis* 40-41
Citrus reticulata 42-43
Cocos nucifera 44-45
Conium maculatum 46-47
Cordyline fruticosa (= *Convallaria fruticosa*, *Cordyline terminalis*, *Dracaena terminalis*) 48-49
Crassula ovata 50-51
Crataegus species 52-53
Curcuma longa (= *Amomum curcuma*, *Curcuma domestica*) 54-55
Dendranthema x*grandiflora* (= *Chrysanthemum* x*morifolium*) 56-57
Dianthus caryophyllus 58-59
Dracaena fragrans (= *Pleomele fragrans*, *Dracaenaderemensis*) 60-61
Dracaena sanderana (*Dracaena sanderiana*) 62-63
Epiphyllum oxypetalum 64-65
Euphorbia pulcherrima 66-67
Euphorbia species 138
Ficus benjamina 68-69
Ficus carica 70-71
Ficus microcarpa 72-73
Ficus religiosa 74-75
Fortunella japonica (*Fortunella margarita*) 76-77
Four leaved clover 78-83
Fuchsia species and hybrids 84-85
Gladiolus species and hybrids 86-87
Heavily fruiting plants 88-89
Hedera helix 90-91
Hordeum vulgare 92-93
Hyacinthus orientalis 94-95
Hypericum species 96-97
Ilex species and hybrids 98-99
Jasminum sambac 100-101
Juniperus species and hybrids 102-103
Lageneria siceraria 104-105
Laurus nobilis 106-107
Lilium species and hybrids 108-109
Lycium barbatum var. *barbatum* 138
Lycium chinense var. *chinense* 138
Lycopersicon esculentum cultivated varieties 88
Mangifera indica 110-111
Medicago lupulina 78
Metrosideros excelsa 30
Michelia x*alba* (= *Magnolia* x*alba*, *Michelia longifolia*) 112-113
Michelia champaca (= *Magnolia champaca*) 114-115
Musa species and hybrids 116-117
Nelumbo nucifera (= *Nelumbo speciosa*, *Nelumbium speciosum*) 118-119
Nuytsia floribunda 30
Nyctanthes arbor-tristis 120-121
Ocimum tenuiflorum (= *Ocimum sanctum*) 122-123
Olea europaea 124-125
Onopordum acanthium 138
Oryza sativa 126-127
Oxalis species or *Oxalis acetosella* 79
Pachira aquatica 128-129
Paeonia cultivars 130-131
Paliurus spina-christi 138
Phoenix dactylifera 132-133
Picea abies 30
Picea pungens 30
Pinus species 134-135
Pinus sylvestris 30
Platycladus orientalis 136-137
Prickly, spiny or thorny plants 138-141
Prunus mume and *prunus salicina* 142-143
Prunus persica cultivars 144-145
Prunus serrulata and *Prunus* x*yedoensis* 146-147
Pseudotsuga menziesii 30
Punica granatum 150-151
Rosa species and hybrids 152-155
Rosmarinus officinalis 156-157
Saccarum officinarum 158-159
Salix caprea 160-161
Salix species and hybrids 162-163

Saraca asoca (= *Saraca indica*) 164-165
Sarcopoterium spinosum 138
Solanum mammosum 166-167
Solanum pseudocapsicum 88
Spondias dulcis (= *Spondias cytherea*) 88
Thuja occidentalis 168-169
Thysanolaena latifolia (= *Thysanolaena agrostis*, *Thysanolaena maxima*) 170-171
Trifolium dubium 80
Trifolium repens 81
Vicia faba 172-173
Viscum album 174-175
Vitis vinifera 176-177
Zamioculcas zamiifolia 178-179
Zantedeschia aethiopica (= *Calla aethiopica*) 180-181
Ziziphus spina-christi 138

scientific family names
Anacardiaceae 110
Annonaceae 18
Apiaceae 46
Apocynaceae 2, 6
Aquifoliaceae 98
Araceae 178, 180
Araliaceae 90
Arecaceae 44, 132
Asteraceae 56
Betulaceae 14
Brassicaeae 22
Bromeliaceae 8
Cactaceae 64
Caryophyllaceae 58
Compositae 56
Crassulaceae 50
Cruciferae 22
Cucurbitaceae 102
Cupressaceae 102, 136, 168
Dicksoniaceae 36
Euphorbiaceae 66
Fabaceae 20, 24, 28, 78, 80, 81, 164, 172
Gramineae 12, 92, 126, 158, 170
Hyacintheaceae 94
Hypericaceae 96
Iridaceae 86
Labiatae 122, 156
Lamiaceae 122, 156
Lauraceae 106
Laxmanniaceae 48
Leguminosae 20, 24, 28, 78, 80, 81, 164, 172
Liliaceae 108
Lythraceae 150
Magnoliaceae 112, 114
Malvaceae 26, 128
Meliceae 10
Moraceae 68, 70, 72, 74
Musaceae 116
Nelumbonaceae 118
Oleaceae 100, 120, 124
Onagraceae 84
Oxalidaceae 79
Paeoniaceae 130
Palmae 44, 132
Pinaceae 134
Poaceae 12, 92, 126, 158, 170
Rosaceae 52, 142, 144, 146, 152
Ruscaceae 60, 62
Rutaceae 4, 38, 40, 42, 76
Salicaceae 160, 162
Santalaceae 174
Solanaceae 166
Theaceae 16
Umbelliferae 46
Vitaceae 176
Zingiberaceae 54

shrubs 2-3, 12-13, 14-15, 16-17, 28-29, 38-39, 40-41, 42-43, 48-49, 52-53, 60-61, 62-63, 64-65, 66-67, 70-71, 76-77, 84-85, 88-89, 98-99, 100-101, 102-103, 106-107, 120-121, 124-125, 130-131, 134-135, 138-141, 150-151, 152-155, 156-157, 160-161, 162-163, 166-167, 174-175

spirit plants 18-19, 52-53, 64-65, 68-69, 72-73, 100-101, 112-113, 114-115, 116-117

subtropical plants 2-3, 4-5, 8-9, 10-11, 12-13, 16-17, 24-25, 26-27, 36-37, 38-39, 40-41, 42-43, 44-45, 48-49, 50-51, 54-55, 56-57, 58-59, 60-61, 62-63, 64-65, 66-67, 68-69, 70-71, 72-73, 74-75, 78-83, 84-85, 86-87, 88-89, 90-91, 92-93, 100-101, 102-103, 104-105, 106-107, 108-109, 110-111, 112-113, 114-115, 116-117, 118-119, 120-121, 122-123, 124-125, 126-127, 128-129, 132-133, 134-135, 136-137, 138-141, 142-143, 150-151, 152-155, 156-157, 158-159, 160-161, 162-163, 164-165, 166-167, 170-171, 172-173, 176-177, 180-181

temperate plants 12-13, 14-15, 16-17, 28-29, 30-35, 46-47, 52-53, 56-57, 58-59, 70-71, 76-77, 78-83, 84-85, 86-87, 88-89, 90-91, 92-93, 94-95, 96-97, 98-99, 102-103, 106-107, 108-109, 118-119, 124-125, 126-127, 130-131, 132-133, 134-135, 136-137, 138-141, 142-143, 144-145, 146-147, 150-151, 152-155, 156-157, 158-159, 160-161, 162-163, 168-169, 170-171, 172-173, 174-175, 176-177, 180-181

trees 4-5, 6-7, 10-11, 12-13, 14-15, 16-17, 18-19, 24-25, 26-27, 28-29, 30-35, 38-39, 40-41, 42-43, 44-45, 52-53, 60-61, 66-67, 68-69, 70-71, 72-73, 74-75, 76-77, 84-85, 88-89, 98-99, 102-103, 106-107, 110-111, 112-113, 114-115, 120-121, 124-125, 128-129, 132-133, 134-135, 138-141, 142-143, 144-145, 146-147, 160-161, 162-163, 164-165, 168-169

tropical plants 2-3, 4-5, 6-7, 8-9, 10-11, 12-13, 18-19, 20-21, 24-25, 26-27, 36-37, 38-39, 40-41, 42-43, 44-45, 48-49, 50-51, 54-55, 60-61, 62-63, 64-65, 66-67, 68-69, 72-73, 74-75, 76-77, 78-83, 88-89, 100-101, 102-103, 104-105, 106-107, 110-111, 112-113, 114-115, 116-117, 118-119, 120-121, 122-123, 124-125, 126-127, 128-129, 132-133, 134-135, 136-137, 138-141, 150-151, 152-155, 158-159, 160-161, 162-163, 164-165, 166-167, 168-169, 170-171, 176-177, 178-179

Acknowledgements

We are very indebted to many people who contributed in one way or another to this book. We are particularly grateful to all the people we interviewed who willingly shared their personal knowledge on how they viewed or utilised auspicious and inauspicious plants based on their traditions, culture, religion and supersititions.

We are also appreciative of the advice from Jack B. Fischer and Benito C. Tan as well as the technical assistance by Tan Kai-xin, Louise Neo, Alvin Francis Lok, Ang Wee Foong, Morgany Thangavelu, Yeo Chow Khoon, Ho Chee Lick, Ma Ming and Koh Swee Jong.

We are also very grateful for the generosity of the photographers/institutions who allowed us the use of their photographs: Abdul Aziz Agil, Aila Ventura, Alfred Wein, Andrew Miller, Angie Ng-Chua, Anna Buxton, Anna Sidiropoulou, Bill Strong, Bob Nichols, Boo Chih Min, Bradley Arnett, Chrissie Jamieson, Christian Guthier, Dana Duncan Seil, Dave Quitoriano, Dave Thomas, David Karp, David Lee, David Wright, Derek Chew, Doug Waylett, Elaine Chan, Elisa Arteaga, Eric Hunt, Erwyn Arizo, Gabriela Bruno, Gertrude Kanu, Gypsy Flores, Henk van der Eijk, Inna Shulman, Jaap Cost Budde, Jana Skornikova, Jeff Vanuga, Joseph Levi, Judy Malley, K. Jane Burpee, Kate Kruger, Keith Weller, Lauren Bansemer, Leslie Vella, Maria Porta, Melvin Bagaforo, Michael Loudon, Neil Gilham, Nicole Hamaker, Pat Knight, Patrick Standish, Peggy Greb, Peter Birch, Peter M. Forster, Phillip Merritt, R.R. Smith, Rafael Mendina, Rita van Deemter, Robert H. Mohlenbrock, Rowena Wood, Ruth Ellison, Sarah Thomas, Scott Bauer, Sing H. Lin, Stephen Ausmus, Steve Baskauf, Steve Hurst, Steven Patch, United States Department of Agriculture-Agriculture Research Service (USDA-ARS), United States Department of Agriculture-Natural Resources Conservation Service (USDA-NRCS) PLANTS Database, USDA-NRCS photo gallery, Vicky Scott, Wee Yeow Chin, Werner Drizhal, Winfield Sterling, Wlodzimierz Lukasik and Yeo Chow Khoon.